D0025289

On
the Internet

On the Internet is an original and stimulating discussion of the promises and dangers of the Internet. Going beyond the hype surrounding the World Wide Web, Hubert Dreyfus, a celebrated writer on philosophy and technology, asks what price we pay when we enter cyberspace and leave our bodies behind.

Drawing on a diverse array of thinkers from Plato to Kierkegaard, *On the Internet* is one of the first books to bring philosophical analysis to such questions as whether the Internet can solve the problems of mass education, and bring humanity to a new level of community. Dreyfus shows the source of the attraction of being a ubiquitous net surfer in Plato's disdain for the body, and how Kierkegaard's criticism of the press and public opinion for its lack of commitment anticipate the dangers of the risk-free interactions made possible by the anonymity of the Web. Drawing on recent studies of the isolation experienced by many Internet users, Dreyfus shows how the Internet deprives users of essential embodied human capacities such as trust, and involvement in shared local concerns. With reference to compelling examples from his own teaching experience, the author also shows that 'interactive' education leaves out the shared moods and risks that make possible the acquisition of expertise.

Concise and accessible, this is essential reading for anyone wishing to think clearly about the role of the Internet in our future.

Hubert L. Dreyfus is Professor of Philosophy in the Graduate School at the University of California at Berkeley. His publications include *What Computers Still Can't Do* and *Being-in-the-World*.

Thinking in Action

Series editors: **Simon Critchley**, University of Essex, and **Richard Kearney**, University College Dublin and Boston College.

Thinking in Action is a major new series that takes philosophy to its public. Each book in the series is written by a major international philosopher or thinker, engages with an important contemporary topic, and is clearly and accessibly written. The series informs and sharpens debate on issues as wide ranging as the Internet, religion, the problem of immigration and refugees and the way we think about science. Punchy, short, and stimulating, **Thinking in Action** is an indispensable starting point for anyone who wants to think seriously about major issues confronting us today.

On Cosmopolitanism and Forgiveness Jacques Derrida
On Immigration and Refugees Michael Dummett
On Science B. K. Ridley
On the Internet Hubert L. Dreyfus
On Religion John D. Caputo
On Belief Slavoj Žižek

On the Meaning of Life John Cottingham
On Stories Richard Kearney
On Authority Paul Ricoeur
On Humour Simon Critchley
On Film Stephen Mulhall

HUBERT L. DREYFUS

On
the Internet

London and New York

First published 2001
by Routledge
11 New Fetter Lane, London EC4P 4EE

Simultaneously published in the USA and Canada
by Routledge
29 West 35th Street, New York, NY 10001

Routledge is an imprint of the Taylor & Francis Group

Typeset in Joanna MT by
RefineCatch Ltd, Bungay, Suffolk
Printed and bound in Great Britain by
TJ International Ltd, Padstow, Cornwall

British Library Cataloguing in Publication Data
A catalogue record for this book is available from the British Library

Library of Congress Cataloging in Publication Data
Dreyfus, Hubert L.
On the internet / Hubert Dreyfus.
p. cm. – (Thinking in action)
Includes bibliographical references and index.
1. Information technology – Social aspects. 2. Internet – Social aspects. 3. Social isolation. I. Title. II. Series.

HM851 .D74 2001
303.48'33 – dc21 00–046010

ISBN 0–415–22806–9 (hbk)
ISBN 0–415–22807–7 (pbk)

For Geneviève,

surfing survivor and Website designer,

who has mastered the worst and best of the Internet.

'Body am I, and soul' – thus speaks the child. And why should one not speak like children?

But the awakened and knowing say: body am I entirely, and nothing else; and soul is only a word for something about the body.

Behind your thoughts and feelings, my brother, there stands a mighty ruler, an unknown sage – whose name is self. In your body he dwells; he is your body.

Friedrich Nietzsche, **Thus Spake Zarathustra**,
trans. W. Kaufmann, New York: Viking Press, 1966, p. 34

The body is our general medium for having a world. Sometimes it is restricted to the actions necessary for the conservation of life, and accordingly it posits around us a biological world; at other times, elaborating upon these primary actions and moving from their literal to a figurative meaning, it manifests through them a core of new significance: this is true of motor skills such as dancing. Sometimes, finally, the meaning aimed at cannot be achieved by the body's natural means; it must then build itself an instrument, and it projects thereby around itself a cultural world.

Maurice Merleau-Ponty, **Phenomenology of Perception**,
trans. C. Smith, London: Routledge & Kegan Paul, 1962, p. 146

Acknowledgements

I'm indebted to many people for their invaluable help: to Alastair Hannay for providing the new Kierkegaard translation I use in Chapter 4; to Gordon Rios and Kenneth Goldberg for their patient explanations of how searching on the Net actually works; to Arun Tripathi for forwarding to me more material on the Internet than anyone could ever have time to read; to Charles Spinosa, Sean Kelly, Béatrice Han, Stuart Dreyfus; to Mark Wrathall for tough objections and help in answering them; and to Jos de Mul and his seminar, 'The Internet and public administration', for working through the whole manuscript and making many helpful critical comments; to Geneviève Dreyfus for teaching me to use the Net and for preparing the final manuscript; and especially to David Blair, whose sophisticated Wittgensteinian understanding of document retrieval, not only helped me understand the problems of information retrieval on the Internet, but also helped me see how these problems fit with my own Merleau-Pontian sense of the limitations of life in cyberspace.

The extract from Bob Dylan's 'Highway 61 Revisited' on p. 73 is reprinted by kind permission of Sony/ATV Music Publishing (UK) Ltd.

Introduction

I teach you the overman. Man is something that is to be overcome. What have you done to overcome him?

Friedrich Nietzsche, **Thus Spake Zarathustra**, prologue

Why seek to become posthuman? . . . Certainly, we can achieve much while remaining human. Yet we can attain higher peaks only by applying our intelligence, determination, and optimism to break out of the human chrysalis. . . . Our bodies . . . restrain our capacities.

Nietzsche citation and response by Max More, founder of the Extropian Institute[1]

http://www.ct.heise.de/tp/english/inhalt/co/2041/1.html

The Internet is not just a new technological innovation; it is a new *type* of technological innovation; one that brings out the very essence of technology. Up to now, technological innovators have generally produced devices that served needs that were already recognized, and then discovered some unexpected side effects. So Alexander Graham Bell thought the telephone would be useful for communication in business but would not be accepted into people's homes, let alone intrude as they walked down the street. Likewise, Henry Ford thought of the automobile as giving people cheap reliable, individualized transportation, but he did not imagine it would destroy the inner cities and liberate adolescent sex. The Net is different. It was originally intended for communication between scientists, but now *that* is a side effect. We have come to realize that the Net is too gigantic and protean for us to think of it as a device for satisfying *any* specific need, and each new use it affords is a surprise. If the essence of technology is to make *everything* easily accessible and optimizable, then the Internet is the perfect technological device. It is the culmination of the same tendency to make everything as flexible as possible that has led us to digitalize and interconnect as much of reality as we can.[2] What the Web will allow us to do is literally unlimited. This pure flexibility naturally leads people to vie for outrageous predictions as to what the Net will become. We are told that, given its new way of linking and accessing information, the Internet will bring a new era of economic prosperity, lead to the development of intelligent search engines that will deliver to us just the information we desire, solve the problems of mass education, put us in touch with all of reality, allow us to have even more flexible identities than we already have and thereby add new dimensions of meaning to our lives.

But, compared with the relative success of e-commerce, the other areas where a new and more fulfilling form of life has been promised have produced a great deal of talk but few happy results.[3] In fact, researchers at Carnegie-Mellon University were surprised to find that, when people were given access to the World Wide Web, they found themselves feeling isolated and depressed. *The New York Times* reports:

> The results of the $1.5 million project ran completely contrary to expectations of the social scientists who designed it and to many of the organizations that financed the study. . . . 'We were shocked by the findings, because they are counterintuitive to what we know about how socially the Internet is being used,' said Robert Kraut, a social psychology professor at Carnegie Mellon's Human Computer Interaction Institute. 'We are not talking here about the extremes. These

were normal adults and their families, and on average, for those who used the Internet most, things got worse.'[4]

The researchers sum up their findings as follows:

> This research examined the social and psychological impact of the Internet on 169 people in seventy-three households during their first one to two years on-line. . . . In this sample, the Internet was used extensively for communication. Nonetheless, greater use of the Internet was associated with declines in participants' communication with family members in the household, declines in the size of their social circle, and increases in their depression and loneliness.[5]

The authors conclude that what is missing is people's actual embodied presence to each other:

> On-line friendships are likely to be more limited than friendships supported by physical proximity. . . . Because on-line friends are not embedded in the same day-to-day environment, they will be less likely to understand the context for conversation, making discussion more difficult and rendering support less applicable. Even strong ties maintained at a distance through electronic communication are likely to be different in kind and perhaps diminished in strength compared with strong ties supported by physical proximity. The interpersonal communication applications currently prevalent on the Internet are either neutral toward strong ties or tend to undercut rather than promote them.[6]

This surprising discovery shows that the Internet user's disembodiment has profound and unexpected effects. Presumably, it affects people in ways that are different from the way most tools do because it can become the main way its users relate to the rest of the world. Given these surprises and disappointments, we would naturally like to know what are the benefits and the dangers of living our lives on-line? Only then might we hope to have a glimmer concerning what the Net can become and what we will become in the process of living through it.

According to the most extreme Net enthusiasts, the long-range promise of the Net is that each of us will soon be able to transcend the limits imposed on us by our body. As John Perry Barlow, one of the foremost proponents of this vision, puts it, the electronic frontier is 'a world that is both everywhere and nowhere, but it is not where bodies live'.[7] By our body, such visionaries seem to mean not only our physical body with its front and back, arms and legs, and ability to move around in the world, but also our moods that make things matter to us, our location in a particular context where we have to cope with things and people, and the many ways we are exposed to disappointment and failure as well as to injury and death. In short, by embodiment they include all aspects of our finitude and vulnerability. In the rest of this book, I will understand the body in these broad terms.

Yeats lamented that his soul was 'fastened to a dying animal' ('Sailing to Byzantium', in *The Tower*, 1928) and it is easy to see the attraction of completing human evolution by leaving behind the animal bodies in which our linguistic and cultural identities are now imprisoned. Who wouldn't wish to become a disembodied being who could be anywhere in the universe and make backup copies of himself to avoid injury and death? Web visionaries and many others would be delighted to be free from deformities, depression, sickness, old age, and death. This is the promise offered us by computer-inspired futurists such as Hans Moravec[8] and Ray Kurzweil.[9] It is typified on the Net (where else?) by such international groups as the Extropians, whose leader, Max More, is quoted in the epigraph to this Introduction. But even more down-to-earth gurus subscribe to the dream that we are entering a new level of civilization. According to industry consultant Esther Dyson, 'Cyberspace is the land of knowledge, and the exploration of that land can be a civilization's truest, highest calling.'[10]

Leaving the body behind would have pleased Plato, who subscribed to the saying that the body was the tomb of the soul[11] and followed Socrates in claiming that it should be a human being's highest goal to 'die to his body' and become a pure mind. As Socrates put it: 'In despising the body and avoiding it, and endeavoring to become independent – the philosopher's soul is ahead of all the rest.'[12] But that makes it surprising that the Extropians claim to be following Nietzsche, not Plato, when they say we should transcend our humanity.

In fact, Nietzsche's anti-Platonic view of the body is in the very book about the overman the Extropians love to quote. In a section called 'On the Despisers of the Body' Nietzsche has Zarathustra say, as if in direct response to the Extropians: 'I shall not go your way, O despisers of the body! You are no bridge to the overman!'[13] And he continues:

> 'I,' you say, and are proud of the word. But greater is that in which you do not wish to have faith – your body and its great reason: that does not say 'I,' but does 'I.' . . . Behind your thoughts and feelings, my brother, there stands a mighty ruler, an unknown sage – whose name is self. In your body he dwells; he is your body.[14]

Nietzsche thought that the most important thing about human beings was not their intellectual capacities but the emotional and intuitive capacities of their body. In his relentless battle against Platonism and Christianity, even in its most hidden forms in science and technology, Nietzsche, indeed, looked forward to our transcending our human limitations and becoming overmen, but by that he meant that human beings, rather than continuing to deny death and finitude, would finally have the strength to affirm their bodies and their mortality.

So the issue we have to face is: can we get along without our bodies? Is the body just a remnant of our descent from the animal – a limitation on our freedom which the human race is now positioned to outgrow, as the Extropians claim – or does the body play a crucial role even in our spiritual and intellectual life, as Nietzsche contends? If Nietzsche is right, the Net's supposed greatest advantage, freedom from the limits imposed by our bodies, is, ironically, its Achilles' heel.

As a philosopher, I'm not going to become involved in condemning some specific uses of the Internet and praising others. My question is a more speculative one: what if the Net became central in our lives? What if it becomes what Joseph Nye, dean of Harvard University's Kennedy School of Government, calls an 'irresistible alternative culture'? To the extent that we came to live a large part of our lives in cyberspace, would we become super- or infra-human?

In seeking an answer, we should remain open to the possibility that, when we enter cyberspace and leave behind our animal-shaped, emotional, intuitive, situated, vulnerable, embodied selves, and thereby gain a remarkable new freedom never before available to human beings, we might, at the same time, necessarily lose some of our crucial capacities: our ability to make sense of things so as to distinguish the relevant from the irrelevant, our sense of the seriousness of success and failure that is necessary for learning, and our need to get a maximum grip on the world that gives us our sense of the reality of things. Furthermore, we would be tempted to avoid the risk of genuine commitment, and so lose our sense of what gives meaning to our lives. Indeed, in what follows, I hope to show that, if our body goes, so does relevance, skill, reality, and meaning. If that is the trade-off, the prospect of living our lives in and through the Web may not be so attractive after all.

SUMMARY

Chapter 1. The limitations of hyperlinks. The hope for intelligent information retrieval, and the failure of AI. How the actual shape and movement of our bodies play a crucial role in our making sense of our world, so that loss of embodiment would lead to *loss of the ability to recognize relevance*.

Chapter 2. The dream of distance learning. The importance of mattering for teaching and learning. Apprenticeship and the need for imitation. Without involvement and presence *we cannot acquire skills*.

Chapter 3. The absence of telepresence. The body as source of our sense of our grip on reality. How the loss of background coping and attunement endemic to telepresence would lead to the *loss of a sense of the reality of people and things*.

Chapter 4. Anonymity and nihilism. Meaning in our lives requires genuine commitment and real commitment requires real risks. The anonymity and safety of virtual commitments on-line would lead to *life without meaning*.

The Hype about Hyperlinks

One

The AI Problem, as it's called – of making machines behave close enough to how humans behave intelligently – . . . has not been solved. Moreover, there is nothing on the horizon that says, I see some light. Words like 'artificial intelligence,' 'intelligent agents,' 'servants' – all these hyped words we hear in the press – are restatements of the mess and the problem we're in.

We would love to have a machine that could go and search the Web, and our personal stores, knowing our preferences, and knowing what we mean when we say something. But we just don't have anything at that level.

Michael Dertouzos, Director, Laboratory for Computer Science, MIT[1]

The Web is vast and growing exuberantly. At a recent count, it had over a billion pages and it continues to grow at the rate of at least a million pages a day.[2] (It is characteristic of the Web that these statistics, as you read them, are already far out of date.) There is an amazing amount of useful information on the Web but it is getting harder and harder to find. The problem arises from the way information is organized (or, better, disorganized) on the Web. The way the Web works, each element of this welter of information is linked to many other elements by hyperlinks. Such links can link any element of information to any other element for any reason that happens to occur to whoever is making the link. No authority or agreed-upon catalogue system constrains the linker's associations.[3]

Hyperlinks have not been introduced because they are more useful for retrieving information than the old hierarchical ordering. Rather, they are the natural way to use the speed and processing power of computers to relate a vast amount of information without needing to understand it or impose any authoritarian or even generally accepted structure on it. But, when everything can be linked to everything else without regard for purpose or meaning, the size of the Web and the arbitrariness of the links make it extremely difficult for people desiring specific information to find the information they seek.

The problem of retrieving relevant information from a corpus of hyperlinked elements is as new as the Net. The traditional way of ordering information depends on someone – a zoologist, a librarian, a philosopher – working out a classification scheme according to the meanings of the terms involved, and the interests of the users.[4] People can then enter new information into this classification scheme on the basis of what they understand to be the meaning of the categories and the new information. If one wants to use the information, one has to depend on those who wrote and used the classifications to have organized the information on the basis of its meaning, so that users can find the information that is relevant given their purposes.

David Blair, Professor of Computer and Information Systems at the University of Michigan,[5] points out that most 'traditional' classification schemes were explicitly or implicitly linked to a 'practice' of some kind. The life-sciences are the obvious example, but there are other less formal practices that form the foundation of such orderings, such as the timeless practice of farming, where the farmer must be able to identify many kinds of plants, animals, pests, diseases, weather conditions, seasons, etc. While some of the links on the Web can be between Websites that concern specific practices, most are not linked to any practice. Without the demands of a practice to constrain what should be linked to what, the links can proliferate wildly.[6]

Since Aristotle, we have been accustomed to organize information in a hierarchy of broader and broader classes, each including the narrower ones beneath it. So we descend from things, to living things, to animals, to mammals, to dogs, to collies, to Lassie. When information is organized in such a hierarchical database, the user can follow out the meaningful links, but the user is forced to commit to a certain class of information before he can view more specific data that fall under that class. For example, I have to commit to an interest in animals before I can find out what I want to know about tortoises; and once having made that commitment to the animal line in the database, I can't then examine the data on problems of infinity without backtracking through the commitments I have made.

When information is organized by hyperlinks, however, as it is on the Web, instead of the relation between a class and its members, the organizing principle is simply the interconnectedness of all elements. There are no hierarchies; everything is linked to everything else on a single level. Thus, hyperlinks allow the user to move directly from one data entry to any other, as long as they are related in at least some tenuous fashion. The whole of the Web lies only a few links away from any page. With a hyperlinked database, the user is encouraged to traverse a vast network of information, all of which is equally accessible and none of which is privileged. So, for instance, among the sites that contain information on tortoises suggested to me by my browser, I might click on the one called 'Tortoises – compared to hares', and be transported instantly to an entry on Zeno's paradox.

One can illustrate the opposition of the old and new way of organizing and retrieving information, and the attraction of each, with a contrast between the old library culture and the new kind of libraries made possible by hyperlinks. The oppositions show the transformation of a meaning-driven, semantic structuring of information into a formal, syntactic structuring, where meaning plays no role. Table 1 shows a systematization of a few of the oppositions.

OLD LIBRARY CULTURE	HYPERLINKED CULTURE
Classification	**Diversification**
a. stable	a. flexible
b. hierarchically organized	b. single-level
c. defined by specific interests	c. allowing all possible associations
Careful selection	**Access to everything**
a. quality of editions	a. inclusiveness of editions
b. authenticity of the text	b. availability of texts
c. eliminate old material	c. save everything
Permanent collections	**Dynamic collections**
a. preservation of a fixed text	a. intertextual evolution
b. interested browsing	b. playful surfing

Table 1: Opposition between old and new systems of information retrieval

Clearly, the user of a hyper-connected library would no longer be a modern subject with a fixed identity who desires a more complete and reliable model of the world,[7] but rather a postmodern, protean being ready to be opened up to ever new horizons. Such a new being is not interested in collecting what is significant but in connecting to as wide a web of information as possible.

Web surfers embrace proliferating information as a contribution to a new form of life in which surprise and wonder are more important than meaning and usefulness. This approach appeals especially to those who like the idea of rejecting hierarchy and authority and who don't have to worry about the practical problem of finding relevant information. So postmodern theorists and artists embrace hyperlinks as a way of freeing us from anonymous specialists organizing our databases and deciding for us what is relevant to what. *Quantity* of connections is valued above any judgement as to the *quality* of those connections. The idea has an all-American democratic ring. As Fareed Zakaria, the managing editor of *Foreign Affairs*, observes: 'The Internet is profoundly disrespectful of tradition, established order and hierarchy, and that is very American.'[8]

Those who want to use the available data, however, have to find the information that is meaningful and relevant to them given their current concerns. But, given that in a hyperlinked database anything may be linked to anything else, this is a very challenging task. Since hyperlinks are made for all sorts of reasons and since there is only one basic type of link, the searcher cannot use the meaning of the links to arrive at the information he is seeking. The problem is that, as far as meaning is concerned, all hyperlinks are alike. As one researcher puts it, the retrieval job is worse than looking for a needle in a haystack; it's like looking for a specific needle in a needle stack. Given the lack of any semantic content determining the connections, any means for searching the Web must be a formal, syntactic technique for manipulating meaningless symbols so as to try to locate relevant, meaningful, semantic content.

The difficulty of using meaningless mechanical operations

to retrieve meaningful information did not await the arrival of the Net. It arises whenever anyone seeks to retrieve information relevant to a specific purpose from a database not organized to serve that particular purpose. In a typical case, researchers may be looking for published papers on a topic they are interested in, but the mere words in the titles of the papers do not enable a search engine to return just those documents or Websites that meet a specific searcher's needs.

To understand the problem it helps to distinguish Data Retrieval (DR) from Information Retrieval (IR). David Blair explains the difference:

> Data Base Management Systems have revolutionised the management and retrieval of data – we can call directory assistance and get the phone number of just about anyone anywhere in the US or Canada; we can walk to an ATM in a city far away from our home town and withdraw cash from our home bank account; we can go to a ticket office in Michigan and buy a reserved seat for a play in San Francisco; etc. All of this is possible, in part, because of the large-scale, reliable database management systems that have been developed over the last 35 years.
>
> Data retrieval operates on entities like 'names,' 'addresses,' 'phone numbers,' 'account balances,' 'social security numbers,' – all items that typically have clear, unambiguous references. But although some of the representations of documents have clear senses and references – like the author or title of a document – many IR searches are not based on authors or titles, but are interested in the 'intellectual content' of the documents (e.g., 'Get me any reports that analyse Central European investment

prospects in service industries']. Descriptions of *intellectual content* are almost never determinate, and on large retrieval systems, especially the WWW, subject descriptions are usually hopelessly imprecise/indeterminate for all but the most general searching.

So searching for a known URL on the WWW is simple and easy; it has the precision and directedness of data retrieval. But searching for a Web page with specific intellectual content using Web search engines can be very difficult, sometimes impossible.[9]

The difference between Data Retrieval and Document Retrieval can be summed up as shown in Table 2.

DATA RETRIEVAL	DOCUMENT RETRIEVAL
1. Direct ('I want to know X')	1. Indirect ('I want to know about X')
2. Necessary relation between a request and a satisfactory answer	2. Probabilistic relation between a request and a satisfactory document
3. Criterion of success = correctness	3. Criterion of success = utility
4. Scaling up is not a major problem	4. Scaling up is a major problem

Table 2: The differences between data retrieval and document retrieval

Before the advent of the Web and Web search engines, the attempted solution to the document retrieval problem was to have human beings – that is, indexers who understood the documents – help describe their contents so that they might be retrieved by those who wanted them. But there simply aren't enough cataloguers to index the Web – it's too large and it's growing too fast.

To understand the magnitude of the access problem, it's helpful to consider an analogy provided by Blair:

> Suppose we wanted to find a book that is one of several hundred accessible to us. This is rather like finding a particular individual in a crowded room of modest size. Not a particularly difficult problem, even if our description of the book or person we are looking for is fairly general. But suppose we wanted to find a book in a small library of 50,000 books. Although we have all been to libraries of this size, it may still be difficult to imagine the magnitude of the task. Consider a similar problem: Many professional baseball parks in the U.S. hold around 50,000 spectators, so we might be able to better visualise our search task if we imagine our goal is to find a single individual attending a sold-out game at, say, Fenway Park. But now our task is more formidable. Suppose also, that our guidelines for finding the person we want are fairly general: that he is middle-aged, has dark hair, dark eyes, is 5′10″ and slim. Now suppose we are searching for a book in a moderately large library of a few 100,000 books. Here, the analogy would be to finding someone at a Rolling Stones concert in New York's Central Park. But even now, we have yet to comprehend the magnitude of the search space on the INTERNET. Searching through the millions of intellectual resources that are currently available through the INTERNET, utilising only the search tools also currently available, is analogous to searching through N.Y. City for a specific person with only the general description that he has dark hair, dark eyes, is middle-aged and slim.[10]

In the face of such a horrendous problem, researchers

working on information retrieval turned to Artificial Intelligence (AI). Since the 1960s, AI researchers have been working to solve the problem of getting computers, which are syntactic engines sensitive only to the form or shape of their input, to behave like human beings who are sensitive to semantics or meaning. So, naturally, researchers turned to AI for help in programming computers to find just those documents whose relevance would have been recognized by a human being conducting a search. At first, AI researchers were optimistic that they could represent all the facts about the world people cared about by representing a few million facts, and adding rules for finding which facts were relevant in any given situation. But in the late 1970s and early 1980s AI researchers reluctantly came to recognize that, in order to produce artificial intelligence, they would have to make explicit and organize the commonsense knowledge people share, and that was a huge task.[11] The most famous proponent of this approach is Douglas Lenat.[12]

Lenat understands that our commonsense knowledge is not the sort of knowledge found in encyclopedias, but, rather, is the sort of knowledge taken for granted by those writing articles in encyclopedias. Such background knowledge is so obvious to us that we hardly ever notice it. Lenat points out that to understand an article about George Washington, for example, we may need to know such facts as that, when he was in the Capitol, so was his left foot, and that, when he died, he stayed dead. So, in 1985, Lenat proposed that, over the next ten years, he would capture this common sense by building 'a single intelligent agent whose knowledge base contains . . . millions of entries'.[13]

Lenat has now spent fifteen years and at least 15 million dollars developing CYC, a commonsense knowledge database, in the attempt to enable computers to understand such commonsense concerns as requests for information. It is supposed to be a first step towards solving the information retrieval problem. To demonstrate the use of CYC, Lenat has developed a photograph retrieval system as an example of how commonsense knowledge plays an essential role in information retrieval. The system is supposed to retrieve on-line images by caption. Instead of a billion images as one might find on the Web, Lenat starts modestly with twenty pictures. A Stanford professor describes his experience with the system as follows:

> The CYC demo was done with 20 images. The request, 'Someone relaxing', yielded one image, 3 men in beachwear holding surfboards. CYC found this image by making a connection between relaxing and previously entered attributes of the image. But even for 20 pictures the system does not work very well.[14]

In so far as this system works at all, it works only because CYC programmers have made explicit as *knowledge* some of the *understanding* we have of relaxation, exercise, effort, and so forth just by having bodies. But most of our understanding of what it's like to be embodied is so pervasive and action-oriented that there is every reason to doubt that it could be made explicit and entered into a database in a disembodied computer.

That, of course, is not a problem for us in our everyday lives. We can find out the answers to questions involving the body by using our body or imagining what it would be like to be doing such and such. So, for example, we understand that pushups are not relaxing, simply by imagining carrying out the activity. But, a picture of someone doing pushups would need to be labelled for CYC by a human programmer as someone making an effort. Only then could CYC 'deduce' that the person was not relaxing.

In general, by having bodies we can generate as needed an indefinitely large number of facts about our bodies, so many that we do not and could not store them all as explicit knowledge. But CYC does not have a body, so, as we have seen, it has to be given all the facts about the body that it needs to know to retrieve information from its database. Moreover, CYC would still not understand how to use the facts it did know to answer some new question involving the body. For example, if one asked CYC if people can chew gum and whistle at the same time, it would have no idea of the answer even if it knew a lot of facts about chewing and whistling, until an embodied human being imagined trying to do it, and then added the answer to CYC's database. But the number of such facts about the body that one would need to make explicit and store because they might be relevant to some request is endless. Happily, by having a body we dispense with the need to store any such facts at all.

But even if all that we understand just by being embodied could be made explicit and entered into CYC's database, there would still be the more general problem of keeping track of which changes in the world required which changes in the database. Even a Website with a straightforward title including the key words 'Bill Clinton' could be on a host of subjects each one of interest only to some sub-set of users. Moreover, which interest the users had would change as the news changed. One day, foreign policy might be the major subject of interest, and the next, the Starr Report. Since what is relevant about Clinton changes from day to day, one would like to have some procedure that would track day-to-day changes in the world so that a computer could update the way the Clinton Websites were organized as the significance of their content changed.

But an indefinitely large number of changes in the world are taking place all the time; the date is changing, and so are the cloud formations, as well as Clinton's weight, age, location, views, etc. Only a few of these changes, however, are relevant to what people on any given day hope to find on the Clinton Website. A procedure for updating the way information is presented on the Web would, therefore, have to be able to ignore almost all the changes taking place in the world and in Clinton's life, and take account only of the relevant ones.

Human beings respond only to the changes that are relevant given their bodies and their interests, so it should be no surprise that no one has been able to program a computer to respond to what is relevant. Indeed, the problem of recognizing which changes are relevant in a given context has been recognized as a serious problem since it showed up in work in Artificial Intelligence in the 1960s. It is called the frame problem, and it remains unsolved to this day.

Lenat realizes that the relevance problem threatens his whole project and proposes, as he must, to replace a sense of relevance based on meaning, by formal axioms. He proposes two kinds of relevance axioms: *specific* and *general*. The idea behind specific relevance axioms is that different sections of the knowledge base 'can be ranked according to their relevance to the problem solving task at hand'.[15] So, for example, if the task given to CYC is to find data relevant to chip design, the program will be guided in its search by an axiom to the effect that the computer section is more relevant than the botany section (although the botany section cannot be ruled out completely, for it might be the source of a useful analogy or two).[16]

But it is not just analogy that makes some facts relevant to other seemingly far-removed facts. Consider the case of a horse race better who knows that a certain jockey has hay fever and, upon observing that the track is covered with golden rod, changes his bet. To get a computer to see that this seemingly irrelevant change in the world is highly relevant to placing a bet is again to face the frame problem. In fact, everything we know can be connected to everything else in a myriad of meaningful ways. In such cases, only an understanding of the meanings involved enables one to select what is relevant to the task in hand. So, rather than helping solve the relevance problem, *specific* relevance axioms just raise the general relevance problem in more dramatic form.

To solve the *general* relevance problem, Lenat proposes general relevance axioms. These are formalizations of such statements as 'It is necessary to consider only events that are temporally close to the time of the event or proposition at issue.'[17] This would bring in the golden rod all right, but, of course, it would bring in an indefinitely large number of other facts about the racetrack, so the relevance problem would not be solved. Moreover, in explaining and defending this axiom, Guha and Levy say 'it is rare that an event occurs and . . . [then after] a considerable period of time . . . suddenly manifests its effects'.[18] But promises and all sorts of health problems, to take just two examples, have exactly the characteristic that relevant effects can be far in the future, and all sorts of historical and psychological facts relevant to my present can be found in my more or less distant past.[19]

When Lenat embarked on his project fifteen years ago, he

claimed that in ten years CYC would be able to read articles in the newspaper and catalogue the new facts it found there in its database without human help. This is the dream of those who expect artificial intelligent agents to find and deliver to each person the information he or she is interested in. But, as Michael Dertouzos points out in the epigraph at the head of this chapter, this breakthrough has not occurred. The moral is, as Don Swanson points out, that 'machines cannot recognize meaning and so cannot duplicate what human judgment in principle can bring to the process of indexing and classifying documents'.[20]

The failure of AI projects such as Lenat's should call our attention to how important our bodies are in making sense of the world. Indeed, our form of life is organized by and for beings embodied like us: creatures with bodies that have hands and feet, insides and outsides; that have to balance in a gravitational field; that move forward more easily than backwards; that get tired; that have to approach objects by traversing the intervening space, overcoming obstacles as they proceed; etc. Our embodied concerns so pervade our world that we don't notice the way our body enables us to make sense of it.[21] We would only notice it by experiencing our disorientation if we were transported to an alien world set up by creatures with radically different – say, spherical or gaseous – bodies, or by observing the helpless confusion of such alien creatures brought into our world.

It would obviously be a great help if we could use our embodied sense of what is relevant for beings with bodies and interests like ours as a background whenever we searched the databases and Websites of the world for relevant information. But, as Lenat's failure to achieve his goal of making explicit our commonsense knowledge has shown, there is no reason to hope we can formalize the understanding we have by virtue of being embodied. So the hope that Artificial Intelligence could solve the relevance problem has now been largely abandoned. There is a vast and ever-growing amount of information out there, and it looks like our only access to it will have to be through computers that don't have bodies, don't share our world, and so don't understand the meaning of our documents and web-sites.

If we leave our embodied commonsense understanding of the world aside, as using computers forces us to do, we have to do things the computer's way and try to locate relevant information by replacing semantics with correlations between formal squiggles. So there is now a whole information retrieval industry devoted to developing Web crawlers and search engines that attempt to approximate a human being's sense of relevance by using only the manipulation of meaningless symbols available to a computer.

Researchers in information retrieval distinguish *recall* and *precision*. In an ideal situation the searcher would retrieve 100 per cent of the relevant documents and 100 per cent of the documents retrieved would be relevant. In short, he would retrieve *all* and *only* the relevant documents. Recall is the percentage of the relevant documents retrieved, while precision is the percentage of retrieved documents that are relevant. Recall and precision are not independent, however, so the searcher is constantly in the difficult position of trading off one for the other. As the searcher tries to maximize recall, precision tends to decrease, and as she tries to maximize precision, recall tends to go down. As a result, a search resulting in 100 per cent recall and 100 per cent precision is, except in rare circumstances, an unattainable ideal.

Recall and precision become even more difficult to maximize as the system gets larger. Given the immense size of the Net, it is estimated that search engines can recall at most 2 per cent of the relevant sites. Blair explains why this important fact is seldom noticed:

> In spite of the size and difficulty searching for specific content, most of the publicity about WWW searching has been positive. IR pioneer Don Swanson observed this phenomenon decades ago, and calls it the 'fallacy of abundance.' The fallacy of abundance is the mistake a searcher makes when he uses a large IR system and is able to find some useful documents. Swanson pointed out that on a sufficiently large system . . . almost *any* query will retrieve some useful documents. The mistake is to think that just because you got *some* useful documents the IR system is performing well. What you don't know is how many *better* documents the system missed.[22]

Indeed, faith in incremental progress towards being able to retrieve just those and only those documents one needs only makes sense if there is one taxonomy, like that of Aristotle or the Dewey decimal system, that captures the way the world is divided up. But in a world of hyperlinks, there can be no saving metaphysical solution.

The early search engines simply created an index of words associated with a list of documents that contained them, with scoring based on whether or not the word was in the title, body, abstract, etc. Researchers generally agree, however, that these techniques have only about a 10 per cent chance of retrieving a useful document for a given query.

So-called popularity engines, which associate pages with specific queries by looking at clicks and time spent on pages, have boosted this number to about 20 per cent. The point of using clicks and time spent on sites to help searchers find what they are looking for is that someone making a request similar to one made by other users would presumably be satisfied by a response that satisfied these other searchers. And satisfaction can be measured by the number of clicks on a certain document and the time spent reading it.

But this does not work as well as hoped. The problem is with the notion of similarity. Everything is similar to everything else in an indefinitely large number of ways – for example, this book and you are similar in that you are near the surface of the earth, made of matter, reflect light, collect dust, etc. – but, we only notice those similarities that matter to us given our bodies and our interests. Since what counts as similar for us depends on our interests, computers cannot make useful judgements of similarity. So we should not be surprised to learn that the method of counting clicks only works if the requests compared are *identical*, and so fails to cover requests that are the same but expressed differently. Moreover, as Gordon Rios reports, 'analysis of large scale query logs (over 100 million and more) show that roughly one half of the queries are unique, so the search engine has no prior click data. This requires aggregating "similar" queries together and leads right back to the problem of similarity.'

The latest technique for finding relevant sites is to replace the responses of search engine users (the clicks) with an analysis of document links. By using the annotations that authors use to link their pages to others, the more advanced search engines have improved the precision of search for some queries. However, the similarity problem again arises because the space of queries is far larger than the exact text used to annotate links. And, of course, link annotations provide a powerful new avenue for spammers; as with click popularity, the spammers have been quick to take advantage of link annotations as a way to boost the ranking of their bogus documents.

A typical problem is that the use of popularity and link descriptions tends to eclipse the less popular pages that may be relevant for a specific context of the query. For example, if you want papers by the famous researcher Michael Jordan, some engines based on click popularity completely ignore him for the basketball player, Michael Jordan. Other engines built for technical audiences fare better for certain users but will not do as well for audiences looking for the player. Clearly the choice of techniques and the documents crawled and installed in a search engine's database determine its point of view in terms of presenting the results to the user.

Gordon Rios, a scientist with Inktomi, the largest provider of search services, sums up the situation as follows:

> We've done large scale internal studies of all the major search engines that suggest that using text, user click behavior, link structure, and annotations provide around 20–30 per cent precision for reasonable queries. We've pushed this a bit further by generating complex statistical models using all these sources of information. Most of us in this industry understand, however, that *we're hitting a wall* in what any system can expect to accomplish.[23]

That 30 per cent is all one can hope for should not come as a surprise. We have seen that there can be no understanding of relevance without commonsense understanding, and no commonsense understanding without a sense of how the world meshes with our embodiment. All search techniques on the Web are crippled in advance by having to approximate a human sense of meaning and relevance grounded in the body, without a body and therefore without commonsense. It follows that search engines are not on a continuum at the far end of which is located the holy grail of computerized retrieval of just that information that is relevant given the interests of the user; they are not even on the relevance dimension.

Don Swanson sums up the point succinctly:

> Consistently effective fully automatic indexing and retrieval is not possible. Our relevance judgments . . . entail knowing who we are, what we are, the kind of world we live in, and why we want what we seek. It is hardly imaginable that a mechanism . . . could acquire such self-knowledge, be given it, or do the job without it.[24]

In cyberspace, then, without our embodied ability to grasp meaning, relevance slips through our non-existent fingers. But, how then do *people* ever find what is relevant to their concerns? This chapter has suggested that, for us, the world is not a meaningless collection of billions of facts. Rather, it is a field of significance organized by and for beings like us with our bodies, desires, interests, and purposes. Not that this solves the mystery of how our brain manages to be tuned to what, at any given moment, is relevant for us, but at least we can see that, given that the world is organized by and for embodied active agents, not by and for disembodied computers, we have a huge head-start in making sense of it and finding the information we want. One thing is sure, as the Web grows, Net users who leave their bodies behind and become dependent on syntactic Web crawlers and search engines will have to be resigned to picking through heaps of junk in the hope of sometimes finding the information they desire.

How Far is Distance Learning from Education?

Two

With knowledge doubling every year or so, 'expertise' now has a shelf life measured in days; everyone must be both learner and teacher; and the sheer challenge of learning can be managed only through a globe-girdling network that links all minds and all knowledge. I call this new wave of technology *hyperlearning*. . . . It is not a single device or process, but a universe of new technologies that both possess and enhance intelligence. The *hyper* in hyperlearning refers not merely to the extraordinary speed and scope of new information technology, but to an unprecedented degree of connectedness of knowledge, experience, media, and brains – both human and nonhuman. . . . We have the technology today to enable virtually anyone who is not severely handicapped to learn anything, at a 'grade A' level, anywhere, anytime.

Lewis J. Perelman, **School's Out**, Avon/Education, 1993, pp. 22–3

Do not spend vast sums of money to buy machinery that you are going to set down on top of existing dysfunctional institutions. The Internet, for example, will not fix your schools. Perhaps the Internet can be part of a much larger and more complicated plan for fixing your schools, but simply installing an Internet connection will almost surely be a waste of money.

Phil Agre, **Telematics and Informatics**, 15(3), 1998, pp. 231–4

In 1922 Thomas Edison predicted that 'the motion picture is destined to revolutionize our educational system and . . . in a few years it will supplant largely, if not entirely, the use of textbooks'. Twenty-three years later, in 1945, William Levenson, the director of the Cleveland public schools' radio station, claimed that 'the time may come when a portable radio receiver will be as common in the classroom as the blackboard'. Forty years after that, the noted psychologist B. F. Skinner, referring to the first days of his 'teaching machines', in the late 1950s and early 1960s, wrote, 'I was soon saying that, with the help of teaching machines and programmed instruction students could learn twice as much in the same time and with the same effort as in a standard classroom.'[1]

For two decades now computers have been touted as a new technology that will revitalize education. In the 1980s they were proposed as tutors, tutees, and drillmasters but none of those ideas seem to have taken hold.[2] Now the latest proposal is that somehow the power of the World Wide Web will make possible a new approach to education for the twenty-first century in which each student will be able to stay at home and yet be taught by great teachers from all over the world.

Many influential people in the United States believe that the development of the Internet will solve the problems of our current educational system.[3] At the secondary school level, we will no longer have to worry about crammed classes, a deficient infrastructure, or the lowering of standards, and, at the college level, we will be able to leave behind the demographic difficulties posed by too many students, limited access to the most expensive universities, and the need for constant retraining as skill requirements change. If the new technology is put to use in the right way, they maintain, a first-class education will be available to everyone, everywhere, in so far as they master the relevant information technology.

The implementation of this vision is well under way. Reed Hundt, who was from 1993 to 1997 Chairman of the Federal Communications Commission presiding over the implementation of the Telecommunications Act of 1996 and helping to negotiate the World Trade Organization Telecommunications Agreement, has no doubts or reservations about the power of the Net to transform education. Indeed, he is euphoric. He boasts that under his guidance:

> the nation began the largest single national program ever to better education from K through 12 – the Snowe-Rockefeller Amendment to the 1996 Telecommunications Act, which at this very moment is causing $4 billion dollars in new money to be spent to put the Internet in every classroom in the country.[4]

And he goes on triumphantly:

> The mayor of Philadelphia told me that this particular feature was the most important thing done by the federal government for education in his lifetime. Rudy Giuliani, told me that this would transform education in New York. I have been told the same thing by all the urban city mayors. In rural areas, the same message is coming forth.

It's a bit hard to see what specific vision of the Net's educational power could generate all this excitement, and Hundt's explanation of what the new connectedness will enable teachers and students to do only adds to the puzzle:

> It is a transformation for education, K through 12. We have always followed the following view. Teachers should be isolated with students. . . . We should never have up-to-date information on any child. . . . It should be impossible to have a dialogue between parents and teachers on how kids are doing. Information should be hoarded, concealed, or destroyed. It should not be created, shared, developed, or learned from. We should make distance learning extremely

> expensive and hard to accomplish, awkward technologically, and economically impossible to implement.

No doubt the Net can change all that, which means that, as far as the public schools are concerned, all one will get for their 4 billion dollars is an efficient e-mail system linking teachers, administrators, and parents, and, for students, access to a lot of on-line information. (Also some kind of distance learning is mentioned but not explained.) But it's hard to see what this transformation in the *method of communication* of those involved in secondary education has to do with what goes on in the classroom. What proposed change in the *method of education* generates all the excitement?

The claims for universities are a lot more specific but, as we shall see, equally irrelevant. Hundt goes on:

> I went back to my old school, Yale, and the dean of one of the professional schools told me: 'Number one, the historic, primary purpose of the university was to have a library so that scholars could gather around it. And second, scholars could meet other scholars and work and talk to them. And third, there would be a validation system so that smart people would be stamped: grade A, Yale; grade A, University of Wisconsin at Madison – whatever. And fourth, it's a place of quiet contemplation.'
>
> All four of these purposes of a university are not just jeopardized but are probably invalid in the information age. No particular reason to go anywhere to have a library when the libraries of the world are available at your fingertips. No particular reason for scholars to actually physically meet with scholars. When you look at the reality in higher education today, the communities of scholars that interact with each other are on the Net; they're not in person anymore. . . . In

terms of validation, how long will a validation system last when fundamentally the Internet disintermediates those systems? And last, in terms of quiet contemplation, it doesn't get any quieter than if you live exactly where you want to live.

So, he said, as far as he could tell, the whole idea of, in his case, Yale University, was threatened.

Now what is striking about this dean's four points as reported by Hundt is that there isn't a word about the role of the university in educating students. Once a university education is defined as a way for scholars to collect information and talk to each other or be left alone, they don't need to be bodily present to each other, and so it looks like the Net could easily replace that vision of the university.

When Hundt attempts to include the students whom the dean seems to have overlooked, he sees them as consumers of information.

The Internet . . . disintermediates everything. And insofar as the university itself is the retailer of knowledge to the consumer, the student, it is disintermediated. Now all that's necessary is for people to be able to trust the new Internet system of education to bring down that old system.

Granted that, insofar as education consists in sending facts from someone who has a lot of information to those who don't have it, the Web works well, but so would videotapes or any recording medium. There must be something more than information-consumption going on in distance learning or there is no point in adding the Internet to the canned lecture. Hundt goes on about empowering the individual scholar, education for all, getting rid of elite universities,[5] but he adds nothing helpful about what education is supposed to be once the Net disintermediates everything.

Of course, many educators hold the opposite view – namely, that colleges and universities are engaged in education, and that education requires face-to-face interaction between teachers and students. For example, Nancy Dye, President of Oberlin College, is sure that 'Learning is a deeply social process that requires time and face-to-face contact. That means professors interacting with students.'[6] Likewise, *The New York Times* reports that 'the American Federation of Teachers . . . critical of the sterility of distance learning, noted, "All our experience as educators tells us that teaching and learning in the shared human spaces of a campus are essential to the undergraduate experience." '[7]

But neither side gives us any reason to accept their pronouncements. In the face of this stand-off with no arguments on either side, we have to take a careful look at education in the light of the new possibilities for distance learning and ask: can distance learning enable students to acquire the skills they need in order to be good citizens skilled in various domains? Or, does learning really require face-to-face engagement, and, if so, why? Just what goes on in classrooms, lecture halls, seminar rooms, and wherever skills are learned?

First, we need to get clear about what skills are and how they are acquired.[8] So, before seeking to evaluate the conflicting claims concerning distance learning, I'll lay out briefly what seem to be the stages in which a student learns by means of instruction, practice, and, finally, apprenticeship, to become an expert in some particular domain and in everyday life. The question then becomes: can these stages be implemented and encouraged on the Web?

STAGE 1: NOVICE

Normally, the instruction process begins with the instructor decomposing the task environment into context-free features that the beginner can recognize without the desired skill. The beginner is then given rules for determining actions on the basis of these features, like a computer following a program.

For purposes of illustration, I'll consider three variations: a motor skill, an intellectual skill, and what takes place in the lecture hall. The student automobile driver learns to recognize such domain-independent features as speed (indicated by the speedometer) and is given rules such as shift to second when the speedometer needle points to ten. The novice chess player learns a numerical value for each type of piece regardless of its position, and the rule: 'Always exchange if the total value of pieces captured exceeds the value of pieces lost.' The player also learns to seek centre control when no advantageous exchanges can be found, and is given a rule defining centre squares and one for calculating extent of control.

In the classroom and lecture hall, the teacher supplies the facts and procedures that need to be learned in order for the student to begin to develop an understanding of some particular domain. The student learns to recognize the features and follow the procedures by drill and practice. Hundt is right that, as long as students are merely consumers of information, as they are at this stage, they don't need to be in a classroom with each other and a teacher at all. Each can learn at his own terminal, wherever and whenever is convenient. Clearly, in this way the Internet can offer an improved version of the correspondence course, but this can't be what the enthusiasts are shouting about.

In any case, merely following rules will produce poor performance in the real world. A car stalls if one shifts too soon on a hill or when the car is heavily loaded; a chess player who always exchanges to gain points is sure to be the victim of a sacrifice by the opponent who gives up valuable pieces to gain a tactical advantage. Understanding a language or a science is much more than memorizing the elements and the rules relating them. The student needs not only the facts but also an understanding of the context in which that information makes sense.

STAGE 2: ADVANCED BEGINNER

As the novice gains experience actually coping with real situations and begins to develop an understanding of the relevant context, he or she begins to note, or an instructor points out, perspicuous examples of meaningful additional aspects of the situation or domain. After seeing a sufficient number of examples, the student learns to recognize these new aspects. Instructional *maxims* can then refer to these new situational *aspects*, recognized on the basis of experience, as well as to the objectively-defined non-situational *features* recognizable by the novice.

The advanced beginner driver uses (situational) engine sounds as well as (non-situational) speed in deciding when to shift. He learns the maxim: shift up when the motor sounds like it's racing and down when it sounds like it's straining. Engine sounds cannot be adequately captured by a list of features. In general, features cannot take the place of a few choice examples in learning the relevant distinctions.

With experience, the chess beginner learns to recognize overextended positions and how to avoid them. Similarly, she begins to recognize such situational aspects of positions as a weakened king's side or a strong pawn structure, despite the lack of precise and situation-free definitions. The player can then follow maxims such as: attack a weakened king's side. Unlike a rule, a maxim requires that one already has some understanding of the domain to which the maxim applies.[9]

At school, mere information is contextualized so that the student can begin to develop an understanding of its significance. The instructor takes on the role of a coach who helps the student pick out and recognize the relevant aspects that organize and make sense of the material. Though aspects can be presented to passive students in front of their terminals, it is more efficient for the student to attempt to use the maxims that have been given, while the instructor points out aspects of the current situation to the student as the student encounters them. Here the teacher needs to be present with the student in the actual situation of thought or action.

Still, at this stage, learning, whether it takes place at a distance or face to face, can be carried on in a detached, analytic frame of mind, as the student follows instructions and is given examples. But to progress further requires a special kind of involvement.

STAGE 3: COMPETENCE

With more experience, the number of potentially relevant elements and procedures that the learner is able to recognize and follow becomes overwhelming. At this point, since a sense of what is important in any particular situation is missing, performance becomes nerve-racking and exhausting, and the student might well wonder how anybody ever masters the skill.

To cope with this overload and to achieve competence, people learn, through instruction or experience, to devise a plan, or choose a perspective, that then determines which elements of the situation or domain must be treated as important and which ones can be ignored. As students learn to restrict themselves to only a few of the vast number of possibly relevant features and aspects, understanding and decision making become easier.

Naturally, to avoid mistakes, the competent performer seeks rules and reasoning procedures to decide which plan or perspective to adopt. But such rules are not as easy to come by as are the rules and maxims given to beginners in manuals and lectures. Indeed, in any skill domain the performer encounters a vast number of situations differing from each other in subtle ways. There are, in fact, more situations than can be named or precisely defined, so no one can prepare for the learner a list of types of possible situations and what to do or look for in each. Students, therefore, must decide for themselves in each situation what plan or perspective to adopt, without being sure that it will turn out to be appropriate.

Given this uncertainty, coping becomes frightening rather than merely exhausting. Prior to this stage, if the rules don't work, the performer, rather than feeling remorse for his mistakes, can rationalize that he hadn't been given adequate rules. But since, at this stage, the result depends on the perspective adopted by the learner, the learner feels responsible for his or her choice. Often, the choice leads to confusion and failure. But sometimes things work out well, and the competent student then experiences a kind of elation unknown to the beginner.

A competent driver leaving the freeway on an off-ramp

curve learns to pay attention to the speed of the car, not whether to shift gears. After taking into account speed, surface condition, criticality of time, etc., he may decide he is going too fast. He then has to decide whether to let up on the accelerator, remove his foot altogether, or step on the brake, and precisely when to perform any of these actions. He is relieved if he gets through the curve without mishap, and shaken if he begins to go into a skid.

The class A chess player, here classed as competent, may decide after studying a position that her opponent has weakened his king's defences so that an attack against the king is a viable goal. If she chooses to attack, she ignores weaknesses in her own position created by the attack, as well as the loss of pieces not essential to the attack. Pieces defending the enemy king become salient. Since pieces not involved in the attack are being lost, the timing of the attack is critical. If she attacks too soon or too late, her pieces will have been lost in vain and she will almost surely lose the game. Successful attacks induce euphoria, while mistakes are felt in the pit of the stomach.

If we were disembodied beings, pure minds free of our messy emotions, our responses to our successes and failures would lack this seriousness and excitement. Like a computer we would have goals and succeed or fail to achieve them, but, as John Haugeland once said of chess machines that have been programmed to win, they seek their goal, but, when it comes to winning, they don't give a damn. For embodied, emotional beings like us, however, success and failure do matter. So the learner is naturally frightened, elated, disappointed, or discouraged by the results of his or her choice of perspective. And, as the competent student becomes more and more emotionally involved in his task, it becomes increasingly difficult for him to draw back and adopt the detached maxim-following stance of the advanced beginner.

But why let learning be infected with all that emotional stress? Haven't we in the West, since the Stoics, and especially since Descartes, learned to make progress by mastering our emotions and being as detached and objective as possible? Wouldn't rational motivations, objective detachment, honest evaluation, and hard work be the best way to acquire expertise?

While it might seem that involvement could only interfere with detached rule-testing, and so would inevitably lead to irrational decisions and inhibit further skill development, in fact, just the opposite seems to be the case. Patricia Benner has studied nurses at each stage of skill acquisition. She finds that, unless the trainee stays emotionally involved and accepts the joy of a job well done, as well as the remorse of mistakes, he or she will not develop further, and will eventually burn out trying to keep track of all the features and aspects, rules and maxims that modern medicine requires. In general, resistance to involvement and risk leads to stagnation and ultimately to boredom and regression.[10]

Since students tend to imitate the teacher as model, teachers can play a crucial role in whether students will withdraw into being disembodied minds or become more and more emotionally involved in the learning situation. If the teacher is detached and computer-like, the students will be too. Conversely, if the teacher shows his involvement in the way he pursues the truth, considers daring hypotheses and interpretations, is open to students' suggestions and objections, and emotionally dwells on the choices that have led him to his conclusions and actions, the students will be more likely to let their own successes and failures matter to them, and rerun the choices that led to these outcomes.

In the classroom and lecture hall the stakes are less dramatic than the risk of having a car accident while driving or of losing an important game of chess. Still, there is the possibility of taking the risk of proposing and defending an idea and finding out whether it fails or flies. If each student is at home in front of his or her terminal, there is no place for such risky involvement. On the contrary, the correspondence-course model of anonymous information consumers, which promoters of distance learning seem to have in mind when they say that the course material will be available to anyone, anywhere, any time, makes such involvement impossible. But, even if we drop the any time, and suppose that the students are all watching the professor at the same time, as with interactive video, and everyone watching hears each student's question, each student is still anonymous and there is still no class before which the student can shine and also risk making a fool of himself. The professor's approving or disapproving response might carry some emotional weight but it would be much less intimidating to offer a comment and get a reaction from the professor if one had never met the professor and was not in her presence. Thus, those who think like President Dye and the American Federation of Teachers may well be right. The Net's limitations where embodiment is concerned – the absence of face-to-face learning – may well leave students stuck at competence.

STAGE 4: PROFICIENCY

Only if the detached, information-consuming stance of the novice, advanced beginner, and distance learner is replaced by involvement, is the student set for further advancement. Then, the resulting positive and negative emotional experiences will strengthen successful responses and inhibit unsuccessful ones, and the performer's theory of the skill, as represented by rules and principles, will gradually be replaced by situational discriminations, accompanied by associated responses. Proficiency seems to develop if, and only if, experience is assimilated in this embodied, atheoretical way. Only then do intuitive reactions replace reasoned responses.

As usual, this can be seen most clearly in cases of action. As the performer acquires the ability to discriminate among a variety of situations, each entered into with involvement, plans are evoked and certain aspects stand out as important without the learner standing back and choosing those plans or deciding to adopt that perspective. Action becomes easier and less stressful as the learner simply sees what needs to be done rather than using a calculative procedure to select one of several possible alternatives. When the goal is simply obvious, rather than the winner of a complex competition, there is less doubt as to whether what one is trying to accomplish is appropriate.

To understand this stage of skill acquisition we must remember that the involved, experienced performer sees goals and salient aspects, but not what to do to achieve these goals. This is inevitable since there are far fewer ways of seeing what is going on than there are ways of reacting. The proficient performer simply has not yet had enough experience with the outcomes of the wide variety of possible responses to each of the situations he can now discriminate, to react automatically. Thus, the proficient performer, after spontaneously seeing the point and the important aspects of the current situation, must still *decide* what to do. And to decide, he must fall back on detached rule- and maxim-following.

The proficient driver, approaching a curve on a rainy day, may *feel in the seat of his pants* that he is going dangerously fast. He must then *decide* whether to apply the brakes or merely to reduce pressure by some specific amount on the accelerator. Valuable time may be lost while making a decision, but the proficient driver is certainly more likely to negotiate the curve safely than the competent driver who spends additional time *considering* the speed, angle of bank, and felt gravitational forces, in order to *decide* whether the car's speed is excessive.

The proficient chess player, who is classed a master, can recognize almost immediately a large repertoire of types of positions. She then deliberates to determine which move will best achieve her goal. She may know, for example, that she should attack, but she must calculate how best to do so.

A student at this level sees the problem that needs to be solved but has to figure out what the answer is.

STAGE 5: EXPERTISE

The *proficient performer*, immersed in the world of his skilful activity, *sees* what needs to be done, but has to *decide* how to do it. The *expert* not only sees what needs to be achieved; thanks to his vast repertoire of situational discriminations, he also sees immediately how to achieve his goal. Thus, the ability to make more subtle and refined discriminations is what distinguishes the expert from the proficient performer. Among many situations, all seen as similar with respect to plan or perspective, the expert has learned to distinguish those situations requiring one reaction from those demanding another. That is, with enough experience in a variety of situations, all seen from the same perspective but requiring different tactical decisions, the brain of the expert gradually decomposes this class of situations into subclasses, each of which requires a specific response. This allows the immediate intuitive situational response that is characteristic of expertise.

The chess Grandmaster experiences a compelling sense both of the issue *and* the best move. Excellent chess players can play at the rate of 5 to 10 seconds a move and even faster without any serious degradation in performance. At this speed they must depend almost entirely on intuition and hardly at all on analysis and comparison of alternatives. It has been estimated that an expert chess player can distinguish roughly 50,000 types of positions. For much expert performance, the number of classes of descriminable situations, built up on the basis of experience, must be comparatively large.

The expert driver not only feels in the seat of his pants when speed is the issue; he knows how to perform the appropriate action without calculating and comparing alternatives. On the off-ramp, his foot simply lifts off the accelerator and applies the appropriate pressure to the brake. What must be done, simply is done. As Aristotle says, the expert 'straightway' does 'the appropriate thing, at the appropriate time, in the appropriate way'.

The student, who has mastered the material, immediately sees the solution to the current problem.

What is the role of the teacher at this stage? A student learns by small random variations on what he is doing, and then checking to see whether or not his performance has improved. Of course, it would be better for learning if these small random variations where not random – if they were sensible deviations. If the learner watches someone who is good at doing something, that could limit the learner's random trials to the more promising ones. So observation and imitation of the activity of an expert can replace a random search for better ways to act. In general, this is the advantage of being an apprentice. Its importance is particularly clear in professional schools.

One thing that professional schools must teach is the way the theory the student has learned can be applied in the real world. One way to accomplish this without apprenticeship is for the school to simulate the surroundings that the students are to function in at a later point in their careers. Business schools provide an instructive example. At American schools of business administration two different modes of thought dominate. One is to be found in the so-called analytical school where most teaching focuses on theory. This type of school rarely produces capable business people who are intuitive experts. The other tradition is based on case studies, where real-life situations are described to the students and discussed. This produces better results.

To become an expert, however, it is not sufficient to have worked through a lot of cases. As we have already seen in discussing the move from competence to proficiency, the cases must matter to the learner. Just as flight simulators work only if the trainee feels the stress and risk of the situation and does not just sit back and try to figure out what to do, for the case method to work, the students must become emotionally involved. So, in a business school case study, the student should not be confronted with objective descriptions of situations, but rather be led to identify with the situation of the senior manager and experience his agonized choices and subsequent joys and disappointments. Provided that they draw in the embodied, emotional student, not just his mind, simulations – especially computer simulations – can be useful. The most reliable way to produce involvement, however, is to require that the student work in the relevant skill domain. So we are back at apprenticeship.

Even where the subject matter is purely theoretical, apprenticeship is necessary. Thus, in the sciences, post-doctoral students work in the laboratory of a successful scientist to learn how their disembodied, theoretical understanding can be brought to bear on the real world. By imitating the master, they learn abilities for which there are no rules, such as how long to persist when the work does not seem to be going well, just how much precision should be sought in each different kind of research situation, and so forth. In order to bring theory into relation with practice, this sort of apprenticeship turns out to be essential.

Even in the humanities where there are no agreed-upon theories, the graduate student needs personal guidance. Thus, she normally becomes a teaching assistant where she can interact with a practising researcher and teacher. The teacher can't help but exhibit a certain style of approaching texts and problems and of asking questions. For example, he may manifest an aggressive style of never admitting he is wrong or a receptive style of soliciting objections and learning from his mistakes. It is their teacher's style more than anything else that the teaching assistants pick up and imitate, even though they usually don't realize that they are doing so. An inspiring teacher like Wittgenstein left several succeeding generations of students not only imitating his style of questioning but even his gestures of puzzlement and desperation.

For passing-on a style, apprenticeship is the only technique available. However, if what the expert produced were clones of his or her own style, apprenticeship would be stultifying. Taking the notion of apprenticeship seriously, one has to ask how, within this framework, new styles and innovative ability can be developed. The training of musicians provides a clue. If you are training to become a performing musician, you have to work with an already recognized master. The apprentice cannot help but imitate the master, because when you admire someone and spend time with them, their style becomes your style. But then the danger is that the apprentice will become merely a copy of the master, while being a virtuoso performing artist requires developing a style of one's own.

Musicians have learned from experience that those who follow one master are not as creative performers as those who have worked sequentially with several.[11] The apprentice, therefore, needs to leave his first master and work with a master with a different style. In fact, he needs to study with several such masters. Journeymen in medieval times, and performing artists even now, when they become good enough to develop a style of their own, travel around and work in various communities of practice. In music, the teachers encourage their students to work with them for a while and then go on to other teachers. Likewise, graduate students usually assist several professors, and young scientists may work in several laboratories.

It is easy for us moderns to misunderstand this need for apprenticeship to several teachers. We tend to think, for example, that the music apprentice needs to go to one master because she is good at fingering, to another because she is good at phrasing, and yet another because she is good at dynamics. That would suggest one could divide a skill into components, which is the wrong way to look at it. Rather, one master has one whole style and another has a wholly different style.[12] Working with several masters destabilizes and confuses the apprentice so that he can no longer simply copy any one master's style and so is forced to begin to develop a style of his own. In so doing he achieves the highest level of skill. Let us call it *mastery*. Such mastery would seem to be out of reach of the distance learner.

STAGE 7: PRACTICAL WISDOM

Not only do people have to acquire skills by imitating the style of experts in specific domains; they have to acquire the style of their culture in order to gain what Aristotle calls practical wisdom. Children begin to learn to be experts in their culture's practices from the moment they come into the world. In this task, they are apprenticed to their parents from the word go.

Our cultural style is so embodied and pervasive that it is generally invisible to us, so it is helpful to contrast our style with some other cultural style and compare how it is learned. Sociologists point out that mothers in different cultures handle their babies in different ways.[13] For example, American mothers tend to place babies in their cribs by putting them on their stomachs, which encourages the babies to move around more. Japanese mothers, contrariwise, put their babies on their backs so they will lie still, lulled by whatever they hear and see. American mothers encourage passionate gesturing and vocalizing, while Japanese mothers are much more soothing and mollifying. In general American mothers situate the baby's body and respond to the baby's actions in such a way as to promote an active and aggressive style of behaviour. Japanese mothers, in contrast, promote a greater passivity and sensitivity to harmony. Thus, what constitutes the American baby as an *American* baby is its style, and what constitutes the Japanese baby as a *Japanese* baby is its quite different style.

The general cultural style determines how the baby encounters himself or herself, other people, and things. Starting with a style, various practices will make sense and become dominant and others will either become subordinate or will be ignored altogether. So, for example, babies never encounter a bare rattle. For an American baby a rattle-thing is encountered as an object to make lots of expressive noise with and to throw on the floor in a wilful way in order to get a parent to pick it up. A Japanese baby may treat a rattle-thing this way more or less by accident, but generally, I suspect, a rattle-thing is encountered as serving a soothing, pacifying function like a Native American rainstick.

Once we see that a style governs how anything can show up *as* anything, we can see that the style of a culture governs not only the babies. The adults in each culture are completely shaped by it too. For example, it should come as no surprise to us, given the sketch of Japanese and American culture already presented, that Japanese adults seek contented, social integration, while American adults are still striving wilfully to satisfy their individual desires. Likewise, the style of enterprises and of political organizations in Japan aims at producing and reinforcing cohesion, loyalty, and consensus, while what is admired by Americans in business and politics is the aggressive energy of a *laissez-faire* system in which everyone strives to express his or her own individuality, and where the state, businesses, or other organizations function to maximize the number of desires that can be satisfied without destructive instability.

Like embodied commonsense understanding, cultural style is too embodied to be captured in a theory, and passed on by talking heads. It is simply passed on silently from body to body, yet it is what makes us human beings and provides the background against which all other learning is possible. It is only by being an apprentice to one's parents and teachers that one gains what Aristotle calls practical wisdom – the general ability to do the appropriate thing, at the appropriate time, in the appropriate way. If we were able to leave our bodies behind and live in cyberspace and chose to do so, nurturing children and passing on one's variation of one's cultural style to them would become impossible.

CONCLUSION

At every stage of skill acquisition beyond the first three, involvement and mattering are essential. Like expert systems following rules and procedures, the immortal detached minds envisaged by futurists like Moravec would at best be competent.[14] Distance learning enthusiasts like Hundt need to realize that only emotional, involved, embodied human beings can become proficient and expert. So, while they are teaching specific skills, teachers must also be incarnating and encouraging involvement. Moreover, learning through apprenticeship requires the presence of experts, and picking up the style of life that we share with others in our culture requires being in the presence of our elders. On this basic level, as Yeats said, 'Man can embody the truth, but he cannot know it.'[15]

When one looks at education in detail – from coaching, to manifesting the necessary involvement, to showing how the theory of a domain can be brought to bear on real situations, to developing one's own style – one can see why the university can't be disintermediated. While the Yale Dean Hundt quotes may well turn out to be superfluous, there is plenty of work left for universities like Yale.

Thus, in so far as we want to teach expertise in particular domains and practical wisdom in life, which we certainly want to do, we finally run up against the most important question a philosopher can ask those who believe in the educational promise of the World Wide Web: can the bodily presence required for acquiring skills in various domains and for acquiring mastery of one's culture be delivered by means of the Internet?

The promise of telepresence holds out hope for a positive answer to this question. If telepresence could enable human beings to be present at a distance in a way that captures all that is essential about bodily presence, then the dream of distance learning at all levels could, in principle, be achieved. But if telepresence cannot deliver the classroom coaching and the lecture-hall presence through which involvement is fostered by committed teachers, as well as the presence to apprentices of masters whose style is manifest on a day-to-day basis so that it can be imitated, distance learning will produce only competence, while expertise and practical wisdom will remain completely out of reach. Hyper-learning would then turn out to be mere hype. So our question becomes: how much presence can telepresence deliver?

Disembodied Telepresence and the Remoteness of the Real

Three

She could see the image of her son, who lived on the other side of the earth, and he could see her. . . . 'What is it, dearest boy?' . . . 'I want you to come and see me.' 'But I can see you!' she exclaimed. 'What more do you want?' . . . 'I see something like you . . . but I do not see you. I hear something like you through this phone, but I do not hear you.' The imponderable bloom, declared by discredited philosophy to be the actual essence of intercourse, was ignored by the machine.

E. M. Forster, 'The Machine Stops'[1]

Artists see far ahead of their time. Thus, just after the turn of the last century, E. M. Forster envisioned and deplored an age in which people would be able to sit in their rooms all their lives, keeping in touch with the world electronically. Now we have almost arrived at this stage of our culture. We can keep up on the latest events in the universe, shop, do research, communicate with our family, friends, and colleagues, meet new people, play games, and control remote robots all without leaving our rooms. When we are engaged in such activities, our bodies seem irrelevant and our minds seem to be present wherever our interest takes us.[2]

As we have seen, some enthusiasts rejoice that, thanks to progress in achieving such telepresence, we are on the way to sloughing off our situated bodies and becoming ubiquitous and, ultimately, immortal. Others worry that if we stay in our rooms and only relate to the world and other people through the Net we will become isolated and depressed, as the Carnegie-Mellon study mentioned in the Introduction seems to confirm.

A more recent and more extensive study at Stanford University confirmed the isolation but did not take up the question of the loneliness and depression. *The New York Times* reports:

> In contrast to the Carnegie-Mellon study, which focused on psychological and emotional issues, the Stanford survey is an effort to provide a broad demographic picture of Internet use and its potential impact on society. . . . Mr. Nie [the survey director] asserted that the Internet was creating a broad new wave of social isolation in the United States, raising the specter of an atomized world without human contact or emotion.[3]

The Stanford researchers, like the sponsors of the Carnegie-Mellon survey, were surprised by their findings. They lament that no one is trying to look ahead to what, if anything, we will lose if we limit ourselves to disembodied interactions. ' "No one is asking the obvious questions about what kind of world we are going to live in when the Internet becomes ubiquitous", Mr. Nie said.'[4] Since that is precisely what we are trying to do here, we had better get on with our work.

Lovers of the Internet claim that we will soon be able to live our lives through a vast Network that will become more and more dense like a tissue or like an invisible ocean in which we will swim. They see this as a great opportunity. *Wired Magazine* tells us:

> Today's metaphor is the network – a vast expanse of nodes strung together with dark, gaping holes in between. But as

> the threads inevitably become more tightly drawn, the mesh will fill out into a fabric, and then – with no voids whatsoever – into an all-pervasive presence, both powerful and unremarkable. . . . In the words of Eric Brewer, a specialist on computer security and parallel computing, it will be 'a giant, largely invisible infrastructure that makes your life better.'[5]

Given that many people now agree that, as things are going, we will soon live our lives through such a vast, invisible, interconnected infrastructure, we must surely ask: *will* it, indeed, make our lives better? What would be gained and what, if anything, would be lost if we were to take leave of our situated bodies in exchange for ubiquitous telepresence in cyberspace? We can break up this question into two: how does relating to *the world* through teletechnology affect our overall sense of reality? And what, if anything, is lost when human beings relate to *each other* by way of teletechnology? To answer these questions, we will first have to explore the more general question: what is telepresence and how is it related to our everyday experience of being in the presence of things and people?

In modernity, we tend to ask how can we ever get out of our inner, private, subjective experience so as to be in the presence of the things and people in the external world? While this seems an important question to us now, it was not always taken seriously. The Greeks thought of human beings as empty heads turned towards the world. St Augustine worked hard to convince people that they had an inner life. In his *Confessions* he goes out of his way to comment on the amazing fact that St Ambrose could read to himself. 'When he read, his eyes scanned the page and his heart explored the meaning, but his voice was silent and his tongue was still.'[6] But the idea that there was an inner world didn't really take hold until early in the seventeenth century when three influences led René Descartes to make the modern distinction between the contents of the mind and the rest of reality.

To begin with, instruments like the telescope and microscope were extending man's perceptual powers, but along with such indirect access came doubts about the reliability of what one seemed to see by means of such prostheses. The church doubted Galileo's report of spots on the sun and, as Ian Hacking tells us, 'even into the 1860s there were serious debates as to whether globules seen through a microscope were artifacts of the instrument or genuine elements of living material (they were artifacts)'.[7]

At the same time, the sense organs themselves were being understood as transducers bringing information to the brain. Descartes pioneered this research with an account of how the eye responded to light and passed the information on to the brain by means of 'the small fibers of the optic nerve'.[8] Likewise, Descartes understood that other nerves brought information about the body to the brain and from there to the mind. Descartes thought that this showed that our access to the world is *indirect*, that is, that things are never directly present to us.

He then went even further and used reports of people with a phantom limb to call into question our seemingly direct experience that we have bodies:

> I have been assured by men whose arm or leg has been amputated that it still seemed to them that they occasionally felt pain in the limb they had lost—thus giving me grounds to think that I could not be quite certain that a pain I

endured was indeed due to the limb in which I seemed to feel it.[9]

So Descartes concluded that we are never present to the world or even to our own bodies but that all that we can directly experience is the content of our own minds. And, indeed, when we engage in philosophical reflection, it seems we have to agree with Descartes. It seems to us that we do not have direct access to the external world but only to our private, subjective experiences.

If this were our true condition, then the mediated information concerning distant objects and people transmitted to us over the Internet as telepresence would be as present as anything could get. But, in response to the Cartesian claim that all our experience of the world is indirect, pragmatists such as William James and John Dewey emphasized that the crucial question is whether our relation to the world is that of a disembodied detached spectator or an involved embodied agent. On their analysis, what gives us our sense of being in direct touch with reality is that we can control events in the world and get perceptual feedback concerning what we have done.

But even this sort of control and feedback is not sufficient to give the controller a sense of direct contact with reality. As long as we are controlling a robot with delayed feedback, such as Ken Goldberg's Telegarden arm[10] or the Mars Sojourner, what we see on the screen will seem to be mediated by our long-distance equipment, and therefore not truly tele-*present*.

There comes a point in interactive robot control, however, where we are able to cope skilfully with things and people in real time. Then, as in laparoscopic-surgery, for example, the doctor feels himself present at the robot site, the way blind people feel themselves present at the end of their cane. But even though interactive control and feedback may give us a sense of being directly in touch with the objects we manipulate, it may still leave us with a vague sense that we are not in touch with reality. Something about the distance still undermines our sense of direct presence.

One might think that what is missing from our experience as we sit safely at home remotely controlling our car, for example, is a constant readiness for risky surprises. To avoid extremely risky situations is precisely why remotely-controlled planet-exploring vehicles and tools for handling radioactive substances were developed in the first place; but, in the everyday world, our bodies are always in potentially risky situations. So, when we are in the real world, not just as minds but as embodied vulnerable human beings, we must constantly be ready for dangerous surprises. Perhaps, when this sense of vulnerability is absent, our whole experience is sensed as unreal, even if, involved in a sort of super-Imax interactive display, we are swaying back and forth as our car careens around dangerous-looking curves. But aren't believers in the triumph of technology such as the Extropians right on this point? Couldn't we develop a technologically-controlled world so tame that being on our guard all the time was no longer necessary? And wouldn't it still seem real?

Maurice Merleau-Ponty has attempted to answer this question, and refute Descartes, by describing just what gives us our sense of the world being directly present to us. He holds that there is a more basic kind of need than the need for safety – a need we can never banish as long as we have bodies. It is the need to get what Merleau-Ponty calls an optimal grip on the world. Merleau-Ponty points out that, when we are looking at something, we tend, without thinking about it, to find the best distance for taking in both the thing as a whole and its different parts. When grasping something, we tend to grab it in such a way as to get the best grip on it. Merleau-Ponty says:

> For each object, as for each picture in an art gallery, there is an optimum distance from which it requires to be seen: . . . at a shorter or greater distance we have a perception blurred through excess or deficiency. We therefore tend towards the maximum of visibility, and seek a better focus as with a microscope.[11]

According to Merleau-Ponty, it is the body that seeks this optimum:

> My body is geared into the world when my perception presents me with a spectacle as varied and as clearly articulated as possible, and when my motor intentions, as they unfold, receive the responses they expect from the world. This maximum sharpness of perception and action points clearly to a perceptual *ground*, a basis of my life, a general setting in which my body can co-exist with the world.[12]

So, perception is motivated by the indeterminacy of experience and our perceptual skills serve to make determin*able* objects sufficiently determin*ate* for us to get an optimal grip on them. Moreover, we wouldn't want to evolve beyond the tendency of our bodies to move so as to get a grip on the world since this tendency is what leads us to organize our experience into the experience of stable objects in the first place. Without our constant sense of the uncertainty and instability of our world and our constant moving to overcome it, we would have no stable world at all.[13]

Not only is each of us an active body coping with things, but, as embodied, we each experience a constant readiness to cope with things in general that goes beyond our readiness to cope with any specific thing. Merleau-Ponty calls this embodied readiness our Urdoxa[14] or 'primordial belief' in the reality of the world. It is what gives us our sense of the direct presence of things. So, for there to be a sense of presence in telepresence, one would not only have to be able to get a grip on things at a distance; one would need to have a sense of the context as soliciting a constant readiness to get a grip on whatever comes along.

This sense of being embedded in a world with which we are set to cope is easiest to see if we contrast our experience of the direct presence of other people with telepresence such as teleconferencing. Researchers developing devices for providing telepresence hope to achieve a greater and greater sense of actually being in the presence of distant people and events by introducing high-resolution television and surround sound, and by adding touch and smell channels. Scientists agree that 'full telepresence requires a transparent display system, high resolution image and wide field of view, a multiplicity of feedback channels (visual as well as aural and tactile information, and even environmental data such as moisture level and air temperature), and a consistency of information between these'.[15] They assume that the more such multi-channel, real-time, interactive coupling teletechnology gives us, the more we will have a sense of the full presence of distant objects and people.

But even such a multi-channel approach may not be sufficient. Two roboticists at Berkeley, John Canny and Eric Paulos, criticize the attempt to break down human–human interaction into a set of context-independent communication channels such as video, audio, haptics, etc. They point out that two human beings conversing face to face depend on a subtle combination of eye movements, head motion, gesture, and posture and so interact in a much richer way than most roboticists realize.[16] Their studies suggest that a holistic sense of embodied interaction may well be crucial to everyday human encounters, and that this *intercorporeality*, as Merleau-Ponty calls it, cannot be captured by adding together 3D images, stereo sound, remote robot control, and so forth.

Just what is missing can best be seen if we return to the question of distance learning. We ended the last chapter by asking whether the presence of the teacher required for full-fledged learning could be captured by telepresence. We are now in a position to suggest an answer to this question. But, rather than looking at the six stages of skill acquisition from the point of view of the learner, we will look at learning from the point of view of the teacher and ask, what, if anything, does the teacher lose in attempting to teach skills at a distance?

If the teacher is only recording videotape, then there is no telepresence at all, and a great deal is surely lost. For example, if risk is important in the learning process, then when the teacher and the class are present together *both* assume a risk that is not there when they are not interacting – the student risks being called on to demonstrate his knowledge of the subject of the lecture, and the teacher risks being asked a question he cannot answer. If this is the case, then it may mean that distance teaching not only may produce poorer learning opportunities, but it may produce poorer teachers.

It's true that we think of teachers teaching students, but it is also the case that in an interactive classroom environment the students teach the teacher. The teacher learns that certain examples do or do not work, that some material has to be presented differently from others, that he was simply wrong about some fact or theory, or even that there was a better way of looking at the whole question. It's been said that a 'good university' is one that has teachers and learners, but that a 'great university' has only learners. If so, passive distance education, by removing the risk in learning and teaching, deprives students and teachers of what is most important, namely, learning how to learn.

The challenging case is live, interactive, video distance learning, although this is not the use of the Web that administrators find cost-effective and therefore attractive. Still, it is the sort of technology that could produce telepresence if anything can. David Blair has given a great deal of thought to his experience both in the presence of students in the classroom and in interactive teleteaching. Here are some of his observations.

> In the first place I am often aware of a lot of things going on in the class in addition to a student actually asking a question or commenting. Sometimes when a student asks a question I can see, peripherally, other students nodding their heads in agreement with the question. This would indicate that the student's question is important to the rest of the class so I will take more care in answering it fully. At the other end of the attention spectrum, I can often see, again, peripherally, when students are bored or sleeping or chatting amongst themselves. This means I may have to pick up the pace of the lecture and try to regain their attention. In teaching students at a distance, I can't control where the camera points and what it zooms in on, the way I control

what attracts my experienced attention when the class is in front of me.

Second, as I lecture, I'm drawn to the point of view that is most comfortable or informative for me – a point of view that may be different from lecture to lecture and even may change during a lecture. Perhaps this is similar to Merleau-Ponty's notion of 'maximum grip'. To find this point of view requires that I be able to move around during the lecture sometimes approaching the students closely, sometimes moving away.

Finally, much of my sense of the immediate presence of the students in a class comes from my ability to make eye contact with them. My experience with the CU-CMe ('see-you-see-me') technology on computers is that you cannot make eye contact over a visual channel, no matter how good the transmission is. To look into another person's eyes, I would have to look straight into the camera but then I would not be able to see the eyes of the other person since, to do that, I would have to turn from the camera to the student's image on the screen. You can look into the camera or look at the screen, but you can't do both.[17]

What is lost, then, in telepresence is the possibility of my controlling my body's movement so as to get a better grip on the world.

What is also lost, even in interactive video, is a sense of the context. In teaching, the context is the mood in the room. In general, mood governs how people make sense of what they are experiencing. Our body is what enables us to be attuned to the mood. Ask yourself, if you were a telespectator at a party, would you be able to share the mood? Whereas, as Heidegger points out, if you are *present* at a party, it is hard to resist sharing the elation or depression of the occasion.[18] Likewise, there is always some shared mood in the classroom and it determines what matters – what is experienced as exciting or boring, salient or marginal, relevant or irrelevant. The right mood keeps students involved by giving them a sense of what is important.

Like a good teacher, Blair is sensitive to the mood in his classroom. He writes:

> As I became more experienced lecturing, I began to have a sense of the class as not just a collection of students but as a whole – as a single entity. I feel that the class as a whole is attentive, or responsive, or not responsive, or friendly, or skeptical, etc. This feeling is not just the sum of certain students who appear this way, but is a kind of general feeling. I can get this feeling without a sense of any individual students exemplifying these characteristics. I don't think that any telecommunications device could enable me to get that feeling when viewing the audience at a distance.

One can, perhaps, get a sense of the importance of the sort of subtle interactions that Blair so aptly describes by considering the fact that people pay around $60 a seat to go to a play, even though they can see a movie for a fifth as much. This obviously has something to do with being in the presence of the actors. Presumably, the actors, like good lecturers, are, at every moment, subtly and largely unconsciously adjusting to the responses of the audience and thereby controlling and intensifying the mood in the theatre. Thus, the co-presence of audience and performer provides the audience with the possibility of direct interaction with the performer, and it seems clear that it is this communication going on between the performers and the audience that brings the show to life. Also the spectator in the theatre can choose whom to zoom in on, while in a film that choice is made by the director. Thus, the theatre spectator is actively involved in what happens in front of him, and this contributes to his sense of being present in the same world as the actors.

This way of looking at the importance of bodily presence raises a new question. Films and CDs are different from plays and concerts but each, in its own way, seems just as gripping as its embodied counterpart. Clearly, some stage actors can learn to act in movies, and some live performers can succeed as studio musicians able to produce an intense effect without any feedback from an audience. It should be possible, then, for a lecturer to use the feedback from the cameras and microphones that show remote students, to involve those students in the lecture, without his needing to manage the mood in the remote rooms. This possibility can't be excluded a priori. We will just have to wait and see if distance education breeds a new brand of teleteachers – teacher-movie-actors who are as effective as the current teacher-live-performers.

Still, if we follow the movie/play comparison to the end, the idea that the teleteacher could equal the powerful effect of a skilled teacher who is present in the same room with her students seems unlikely. Without the sense of the mood in the room as well as the shared risk, the involvement of the students with a movie-actor teacher will almost surely be less intense than that of students and teachers reacting to each other's presence. So, it seems that, given the skill model I proposed at the beginning of this chapter, in the domain of education at least, each technological advance that makes teaching more economical and more flexible, by making the teacher and student less immediately present to each other, makes the teaching less effective. One would expect to see a decline in involvement and effectiveness, from tutorial teaching to classroom teaching, to large lecture halls, to interactive video, to asynchronous Net-based courses.

Given this trade-off of economy and efficacy, it looks like we might well end up with a two-tiered educational system where those who can afford it will pay five times as much as the distance learning students pay, in order to be in the presence of their professors. This would amount to an elitism not much different from the English elitism of Oxford and Cambridge *vis-à-vis* the other universities that don't have tutorials – the very elitism that, according to Hundt, the democratic levelling produced by distance learning is supposed to eliminate.

The inferiority of distance learning at the college level seems clear, but what about the vocational and postgraduate teaching which is thought to be the forte of the Internet? One study of the advantages of continuous education on the Internet typifies the jargon and the misplaced optimism characteristic of the field.

> Distributed education encompasses distance education but reaches further to imagine a global disaggregation of instructional resources into modular components of excellence which can be reassembled by any organization in the 'business' of certifying quality-assured learning accomplishment (certificates and degrees). The result should be a conveniently and affordably accessible, enriched educational environment that integrates the networked delivery of learningware and asynchronous and synchronous conversations within learning communities of student apprentices, their expert mentors, and their educational and career advisors.[19]

Such claims completely miss the point of mentoring and apprenticeship. As we have already seen, the role of the master is to pass on to the apprentice the ability to apply the theory of some domain in the real world. But, one might well ask, why not just record the master at work and transmit his image to his teleapprentices? For example, why not just put a camera on the head of a doctor teaching interns on his rounds and wire him with a microphone so that the teleinterns can see and hear just what the doctor and the interns who are present see and hear?

What, if anything, would the teleinterns miss? The answer again is immersion in the context. A camera fixed to the doctor's forehead would, indeed, look wherever he focused his attention, so the teleinterns might well see even better than those actually present in the hospital what the doctor was currently seeing. But the problem is that it is the doctor's responsiveness to the whole situation that determines which details he pays attention to and zooms in on. The camera on the doctor's head would, thus, show distant students exactly what feature of the patient's condition the doctor was seeing, but not the background that led that feature to stand out for the doctor so that he zoomed in on it. The teleintern would surely learn something from a televised image of what the doctor pays attention to, but he or she would always remain a prisoner of the doctor's attention setting, just as in a telelecture the professor is a prisoner of the camera operator and the sound engineer in the distant lecture hall. Yet the ability to zoom in on what is significant is one of the most important skills the intern diagnostician has to learn.

So why not also have a camera and microphone that record and transmit the whole ambient hospital scene? The distance-intern could then watch, on a split screen, both the background and what the doctor focuses attention on, and so learn to notice those features of the overall scene that solicit the doctor's attention.

Here, as in the lecture-hall case, the devil is in the phenomenological details. For the doctor who is actually involved in the situation, it's not as if he had two views – one, a wide-angle view of the uninterpreted situation, and the other, a close-up of the details he is focused on. In becoming a diagnostic master, the doctor has learned to see an already-interpreted situation where certain features and aspects spontaneously stand out as meaningful, just as, as one becomes familiar with a strange city, it ceases to look like a jumble of buildings and streets and develops what Merleau-Ponty calls a familiar physiognomy. The intern is trying, among other things, precisely to acquire the doctor's physiognomic perceptual understanding.

So why, if the intern sees the correlation between the uninterpreted scene on half the screen and the relevant features on the other, couldn't he acquire the doctor's physiognomic understanding? Precisely because the technology deprives the learner of bodily involvement in a risky real environment where he has to interpret the scene himself and learn from his mistakes. Merleau-Ponty would argue that, if one does not have the experience of zooming in on the details that, on the basis of previous experience, come to elicit one's attention, and then discovering the hard way when one is right and when one is mistaken as to the relevant details, one will not find that the scene becomes more and more full of meaning. Thus, the distance-apprentice will not learn to respond to the overall scene by being drawn to zoom in on what is significant. But this is precisely what the intern must learn if he is to become an expert diagnostician.

In the real learning situation, where the patient, the doctor, and the interns are directly present, the apprentice doctors can shift their attention to new details they take to be significant and then find out whether they were right or missed something important. If they are thus involved, then, with every success and failure, the overall organization of their background changes, so that in future encounters a different aspect will stand out as significant. There is thus a constantly enriched interaction between the details and the overall significance of the situation. Merleau-Ponty calls this kind of feedback between one's actions and the perceptual world, the intentional arc.[20] And he points out that it functions only if the perceiver is using his body as an 'I can', that is, in this case, if he controls where he looks.

So, to learn to see what the doctor sees, the tele-intern must be able to control the direction each camera points and how much each camera zooms in or out. After all, simply by having a great deal of passive experience, by watching football games on TV, for example, one can become competent at following the ball and even predicting and interpreting the plays. So one might well think that adding control of where one looked would enable the tele-student to acquire an expert feel for any skill domain. In such an ideal distance-learning setup, would anything required for learning be left out?

As we saw in Chapter 2, the learner becomes an expert by reacting to specific situations, and taking to heart the results. On the basis of sufficient such experience, the brain of the beginner gradually comes to connect perception and action so that, in a situation similar to one that has already been experienced, the agent immediately makes a response similar to the response that worked the last time the learner was in that type of situation. But this requires that the learning situations in which one acquires a skill be sufficiently similar to actual situations so that the responses one learns in training carry over to the real world.

So, any form of telelearning, whether interactive or not, must face a final challenge. Can telepresence reproduce the sense of being in the situation so that what is learned transfers to the real world? Experienced teachers and phenomenologists agree that the answer is 'no'. To see in a stark and extreme form the sort of embodied presence any attempt to transmit full presence cannot capture, it helps to take an example from a physical sport like football.

Barry Lamb, Safeties Coach for the Brigham Young University Football Team and a former All-American linebacker and defensive end at Santa Barbara CC (1973–74), reports the following:

> Our players can learn a great deal by watching films, but only to a point. It's hard to say exactly what it is that you can't learn by watching film, but a good player learns to sense the overall situation and to do things instinctively that just don't make sense if you're only looking at what you can see on film. Most game film, of course, is not taken from a player's perspective. But even if you could correct for that, the depth of field is never the same on film as it is in real life.[21] That means that you can't really learn to see the playing field in the right way, or get a feel for the tempo of the game. In addition, there is more to learning how to see a play develop than just having your head or eyes pointed in the right direction. Our players need to learn how to use their peripheral vision to get a feel for what is going on around them, and what your peripheral vision tells you makes you see what is going on in front of you

differently.[22] Moreover, the emotions of the game change how a player sees the field, and those aren't things that one can get a feel for from the film.

Another way to see how the film is too sterile to teach everything our players need to learn is by noticing that opposing players aren't threatening on film in the same way that they are in real life. The fact that there are eleven players in front of you who want to hurt you really makes you see and understand things differently.

In sum, learning to do the right thing, a thing that sometimes doesn't make sense, is something that can only happen when a person experiences a present situation over and over again, whether in practice or in real life.[23]

All this suggests that distance-learners looking at a surround screen and hearing stereo sound would be able to develop a degree of competence. Thus, an intern could become competent at recognizing and, perhaps, even anticipating many of the symptoms the doctor has pointed out, just as an avid TV viewer can learn to recognize and anticipate many of the plays on the footfall field. Furthermore, if the learner could view the scene transmitted by cameras placed exactly where the actual embodied learner would normally be placed, he might even be able to become proficient. But such distance-learners would still lack the experience that comes from responding directly to the risky and perceptually rich situations that the world presents. Without an experience of their embodied successes and failures in actual situations, such learners would not be able to acquire the ability of an expert who responds immediately to present situations in a masterful way. So we must conclude that expertise cannot be acquired in disembodied cyberspace. Distance-learning enthusiasts notwithstanding, apprenticeship can only take place in the shared situations of the home, the hospital, the playing field, the laboratory, and the production sites of crafts. Distance-apprenticeship is an oxymoron.

Once we see that there is a way of being directly present to things and people that is denied by Descartes and all of modern philosophy, we see that there may well be basic limitations on telepresence that go far beyond the problems of distance teaching. Where the presence of people rather than objects is concerned, we sense a crucial difference between those we have access to through our distance senses of hearing, sight, etc. and the full-bodied presence that is literally within arm's reach. This full-bodied presence is more than the feeling that I am present at the site of a robot arm I am controlling from a distance through real-time interaction. Nor is it just a question of giving robots surface sensors so that, through them as prostheses, we can touch other people at a distance. Even the most gentle person–robot interaction would never be a caress, nor could one successfully use a delicately controlled and touch-sensitive robot arm to give one's kid a hug. Whatever hugs do for people, I'm quite sure telehugs won't do it. And any act of intimacy mediated by any sort of robot prosthesis would surely be equally grotesque, if not obscene. Even if our teletechnology goes beyond the imagination of E. M. Forster so that eventually we can use remote-controlled robotic arms and hands to touch other people, I doubt that people could get a sense of how much to trust each other even if they could stare into each other's eyes on their respective screens, while, at the same time, using their robot arms to shake each other's robotic hands.

Perhaps, one day, we will stop missing this kind of bodily contact, and touching another person will be considered rude or disgusting. E. M. Forster envisions such a future in his story:

> When Vashti swerved away from the sunbeams with a cry [the flight attendant] behaved barbarically – she put out her hand to steady her. 'How dare you!' exclaimed the passenger, 'you forget yourself!' The woman was confused, and apologized for not having let her fall. People never touched one another. The custom had become obsolete, owing to the Machine.[24]

For the time being, however, investment bankers know that in order to get two CEOs to trust one another enough to merge their companies, it is not sufficient that they have many teleconferences. They must live together for several days interacting in a shared environment, and it is quite likely that they will finally make their deal over dinner.[25]

Of course, there are many kinds of trust, and the trust that we have that our mail carrier will deliver our mail does not require looking her in the eye or shaking her hand. The kind of trust that requires such body contact is our trust that someone will act sympathetically to our interests even when so doing might go against his or her own.[26]

What is the connection between such trust and embodied presence? Perhaps our sense of trust must draw on the sense of security and well-being each of us presumably experienced as babies in our caretaker's arms.[27] Our sense of reality, then, would not be just the readiness for flight of a hunted animal; it could also be the feeling of joy and security of being cared for. If so, even the most sophisticated forms of telepresence may well seem remote and even obscene if not in some way connected with our sense of the warm, encircling, nearness of a flesh-and-blood human body.

Furthermore, it seems that to trust someone you have to

make yourself vulnerable to him or her and they have to be vulnerable to you. Part of trust is based on the experience that the other does not take advantage of one's vulnerability. I have to be in the same room with someone and know they could physically hurt me or publicly humiliate me and observe that they do not do so, in order to feel I can trust them and make myself vulnerable to them in other ways.

There is no doubt that telepresence can provide some sense of trust, but it seems to be a much-attenuated sense. Perhaps in the future world of the Internet we will none the less come to prefer telepresence to total isolation, like Harlow's monkeys who, lacking a real mother, shun the wire 'mother' and cling desperately to the terry-cloth one – never knowing the comfort and security of a real mother's arms.[28]

Not that we automatically trust anyone who hugs us. Far from it. Just as for Merleau-Ponty it is only on the background of our embodied faith in the presence and reality of the perceptual world that we can doubt the reality of any specific perceptual experience, so we seem to have a background predisposition to trust those who touch us tenderly, and it is only on the basis of this *Urtrust* that we can be mistrustful in any specific case. If that background trust were missing, as it would necessarily be in cyberspace, we might tend to be suspicious of the trustworthiness of every social interaction and withhold our trust until we could confirm its justification. Such a scepticism would complicate if not poison all human interaction.

CONCLUSION

We have now seen that our sense of the reality of things and people and our ability to interact effectively with them depend on the way our body works silently in the background. Its ability to get a grip on things provides our sense of the reality of what we are doing and are ready to do; this, in turn, gives us a sense both of our power and of our vulnerability to the risky reality of the physical world. Furthermore, the body's ability to zero in on what is significant, and then preserve that understanding in our background awareness, enables us to perceive more and more refined situations and respond more and more skilfully; its sensitivity to mood opens up our shared social situation and makes people and things matter to us; and its tendency to respond positively to direct engagement with other bodies; underlies our sense of trust and so sustains our interpersonal world. All this our body does so effortlessly, pervasively, and successfully that it is hardly noticed. That is why it is so easy to think that in cyberspace we could get along without it, and why it would, in fact, be impossible to do so.

Nihilism on the Information Highway: Anonymity vs. Commitment in the Present Age

Four

Oh God said to Abraham, 'Kill me a son' . . .
Well Abe says, 'Where do you want this killin' done?'
God says, 'Out on Highway 61'.
Well Mack the Finger said to Louie the King
I got forty red white and blue shoe strings
And a thousand telephones that don't ring
Do you know where I can get rid of these things
And Louie the King said let me think for a minute son.
And he said yes I think it can be easily done
Just take everything down to Highway 61.
  Now the rovin' gambler he was very bored
He was tryin' to create a next world war
He found a promoter who nearly fell off the floor
He said I never engaged in this kind of thing before
But yes I think it can be very easily done
We'll just put some bleachers out in the sun
And have it on Highway 61.

Bob Dylan, 'Highway 61 Revisited'

In the section of *A Literary Review*, written in 1846, entitled 'The Present Age'[1] Kierkegaard warns that his age is characterized by a disinterested reflection and curiosity that level all differences of status and value. In his terms, this detached reflection levels all qualitative distinctions. Everything is equal in that nothing matters enough that one would be willing to die for it. Nietzsche gave this modern condition a name; he called it nihilism.

Kierkegaard blames this levelling on what he calls the

Public. He says that 'For levelling properly to come about a phantom must first be provided, its spirit, a monstrous abstraction, an all-encompassing something that is a nothing, a mirage – this phantom is the *public*.'[2] But the real villain behind the Public, Kierkegaard claims, is the Press. He warned that 'Europe will come to a standstill at the Press and remain at a standstill as a reminder that the human race has invented something which will eventually overpower it',[3] and he adds: 'Even if my life had no other significance, I am satisfied with having discovered the absolutely demoralizing existence of the daily press.'[4]

But why blame levelling on the public rather than on democracy, technology, or loss of respect for tradition, to name a few candidates? And why this monomaniacal demonizing of the press? Kierkegaard says in his journals that 'Actually it is the Press, more specifically the daily newspaper . . . which make[s] Christianity impossible.'[5] This is an amazing claim. Clearly, Kierkegaard saw the press as a unique cultural/religious threat, but it will take a little while to explain why.

It is no accident that, writing in 1846, Kierkegaard chose to attack the public and the press. To understand why he did so, we have to begin a century earlier. In *The Structural Transformation of the Public Sphere*[6] Jürgen Habermas locates the beginning of what he calls the *public sphere* in the middle of the eighteenth century. He explains that at that time the press and coffeehouses became the locus of a new form of political discussion. This new sphere of discourse was radically different from the ancient polis or republic; the modern public sphere understood itself as being outside political power. This extra-political status was not just defined negatively, as a lack of political power, but seen positively. Just because public opinion is not an exercise of political power, it is protected from any partisan spirit. Enlightenment intellectuals saw the public sphere as a space in which the rational, disinterested reflection that should guide government and human life could be institutionalized and refined. Such disengaged discussion came to be seen as an essential feature of a free society. As the press extended public debate to a wider and wider readership of ordinary citizens, Burke exalted that, 'in a free country, every man thinks he has a concern in all public matters'.[7]

Over the next century, thanks to the expansion of the daily press, the public sphere became increasingly democratized until this democratization had a surprising result which, according to Habermas, 'altered [the] social preconditions of "public opinion" around the middle of the [nineteenth] century.'[8] '[As] the Public was expanded . . . by the proliferation of the Press . . . the reign of public opinion appeared as the reign of the many and mediocre'.[9] Many people, including J. S. Mill and Alexis de Tocqueville, feared 'the tyranny of public opinion,'[10] and Mill felt called upon to protect 'nonconformists from the grip of the Public itself'.[11] According to Habermas, Tocqueville insisted that 'education and powerful citizens were supposed to form an *elite public* whose critical debate determined public opinion'.[12]

'The Present Age' shows just how original Kierkegaard was. While Tocqueville and Mill claimed that the masses needed elite philosophical leadership, and while Habermas agrees with them that what happened around 1850 with the democratization of the public sphere by the daily press is an unfortunate decline into conformism from which the public sphere must be rescued, Kierkegaard sees the public sphere as a new and dangerous cultural phenomenon in which the nihilism produced by the press brings out something that was deeply wrong with the Enlightenment idea of detached reflection from the start. Thus, while Habermas is concerned to recapture the moral and political virtues of the public sphere, Kierkegaard warns that there is no way to salvage the public sphere since, unlike concrete and committed groups, it was from the start the source of levelling.

This levelling was produced in several ways. First, the new massive distribution of desituated information was making every sort of information immediately available to anyone, thereby producing a desituated, detached spectator. Thus, the new power of the press to disseminate information to everyone in a nation led its readers to transcend their local, personal involvement and overcome their reticence about what didn't directly concern them. As Burke had noted with joy, the press encouraged everyone to develop an opinion about everything. This is seen by Habermas as a triumph of democratization, but Kierkegaard saw that the public sphere was destined to become a detached world in which everyone had an opinion about and commented on all public matters without needing any first-hand experience and without having or wanting any responsibility.

The press and its decadent descendant, the talk show, are bad enough, but this demoralizing effect was not Kierkegaard's main concern. For Kierkegaard, the deeper danger is just what Habermas applauds about the public sphere, namely, as Kierkegaard puts it, 'the public . . . eats up all individuality's relativity and concreteness'.[13] The public sphere thus promotes ubiquitous commentators who deliberately detach themselves from the local practices out of which specific issues grow and in terms of which these issues must be resolved through some sort of committed action. What seems a virtue to detached Enlightenment reason, therefore, looks like a disastrous drawback to Kierkegaard. Even the most conscientious commentators don't have to have first-hand experience or take a concrete stand. Rather, as Kierkegaard complains, they justify their views by citing principles. Since the conclusions such abstract reasoning reaches are not grounded in the local practices, its proposals would presumably not enlist the commitment of the people involved, and consequently would not work even if enacted as laws. As Kierkegaard puts it in 'The Present Age':

> The public is not a people, a generation, one's era, not a community, an association, nor these particular persons, for all these are only what they are by virtue of what is concrete. *Not a single one of those who belong to the public has an essential engagement in anything*'.[14]

More basically still, that the public sphere lies outside of political power meant, for Kierkegaard, that one could hold an opinion on anything without having to act on it. He notes with disapproval that '[the public's] artifice, its good sense, its virtuosity consists in letting matters reach a verdict and a decision without ever acting'.[15] This opens up the possibility of endless reflection. For, if there is no need for decision and action, one can look at all things from all sides and always find some new perspective. The accumulation of information thus postpones decision indefinitely since, as one finds out more, it is always possible that one's picture of the world, and, therefore, of what one should do, will have to be revised. Kierkegaard saw that, when everything is up for endless critical commentary, action can always be postponed. '[R]eflection is able at any moment to put things in a new light and allow one some measure of escape'.[16] Thus one need never act.

All that a reflective age like ours produces is more and more knowledge. As Kierkegaard puts it, 'One can say in general of a passionless but reflective age, compared to a passionate one, that it *gains in extensity what it loses in intensity*.'[17] He adds: 'we all know what path to take and what paths can be taken, but no one will take them.'[18] No one stands behind the views the public holds, so no one is willing to act. He wrote in his journal: 'here . . . are the two most dreadful calamities which really are the principle powers of impersonality – the Press and anonymity.'[19] Therefore, the motto Kierkegaard suggested for the press was: 'Here men are demoralized in the shortest possible time on the largest possible scale, at the cheapest possible price.'[20]

In 'The Present Age' Kierkegaard succinctly sums up his view of the relation of the press, the public sphere, and the levelling going on in his time. The desituated and anonymous press and the lack of passion or commitment in our reflective age combine to produce the public, the agent of the nihilistic levelling:

> The abstraction of the press (for a newspaper, a journal, is no political concretion and only an individual in an abstract sense), combined with the passionlessness and reflectiveness of the age, gives birth to that abstraction's phantom, the public, which is the real leveller.[21]

Kierkegaard would surely have seen in the Internet, with its Websites full of anonymous information from all over the world and its interest groups that anyone in the world can join without qualifications and where one can discuss any topic endlessly without consequences, the hi-tech synthesis of the worst features of the newspaper and the coffeehouse.[22] Indeed, thanks to the Internet, Burke's dream has been realized. In news groups, anyone, anywhere, any time, can have an opinion on anything. All are only too eager to respond to the equally deracinated opinions of other anonymous amateurs who post their views from nowhere. Such commentators do not take a stand on the issues they speak about. Indeed, the very ubiquity of the Net tends to make any such local stand seem irrelevant.

What Kierkegaard envisaged as a consequence of the press's indiscriminate and uncommitted coverage is now fully realized on the World Wide Web. Thanks to hyperlinks, meaningful differences have, indeed, been levelled. Relevance and significance have disappeared. And this is an important part of the attraction of the Web. Nothing is too trivial to be included. Nothing is so important that it demands a special place. In his religious writing Kierkegaard criticized the implicit nihilism in the idea that God is equally concerned with the salvation of a sinner and the fall of a sparrow,[23] that 'for God there is nothing significant and nothing insignificant'.[24] He said such a thought would lead one 'to the verge of despair'.[25] On the Web, the attraction and the danger are that everyone can take this godlike point of view. One can view a coffee pot in Cambridge, or the latest super-nova, study the Kyoto Protocol, find out what fellowships are available to a person with one's profile, or direct a robot to plant and water a seed in Austria, not to mention plough through thousands of ads, all with equal ease and equal lack of any sense of what is important. The highly significant and the absolutely trivial are laid out together on the information highway in just the way Abraham's sacrifice of Isaac, red, white and blue shoe strings, a thousand telephones that don't ring, and the next world war are laid out on Dylan's nihilistic Highway 61.

Kierkegaard even foresaw that the ultimate activity the Internet would encourage would be speculation on how big it is, how much bigger it will get, and what, if anything, all this means for our culture. This sort of discussion is, of course, in danger of becoming part of the very cloud of anonymous speculation Kierkegaard abhorred. Ever sensitive to his own position as a speaker, Kierkegaard concluded his analysis of the dangers of the present age and his dark predictions of what was ahead for Europe with the ironic remark that: 'And since in this age, in which so little is actually done, such an extraordinary amount is done in the way of prophecies, apocalypses, hints, and insights into the future, there is probably nothing else for it but to join in'.[26]

The only alternative Kierkegaard saw to the public's levelling and paralyzing reflection was for one to plunge into some kind of activity – any activity – as long as one threw oneself into it with passionate commitment. In 'The Present Age' he exhorts his contemporaries to make such a leap:

> There is as little action and decision these days as shallow-water paddlers having a dangerous desire to swim. But just as the adult being tossed about delightedly by the waves calls to the younger person: 'Come on, just jump right in' – so the decision so to speak lies within existence (but in the individual, naturally) and shouts to the younger person not yet exhausted by an excess of reflection . . .: 'Come on, jump boldly in.' Even if it were a reckless leap, so long as it is decisive – if you have it in you to be a man, then the danger and life's stern judgment upon your recklessness will help you become one.[27]

Such a light-hearted leap out of the shallow, levelled present age into the deeper water is typified for Kierkegaard by people who leap into what he calls the *aesthetic sphere of existence*. Each sphere of existence, as we shall see, represents a way of trying to get out of the levelling of the present age by making some way of life absolute.[28] In the aesthetic sphere, people make enjoyment the centre of their lives.

Such an aesthetic response is characteristic of the Net-surfer for whom information gathering has become a way of life. Such a surfer is curious about everything and ready to spend every free moment visiting the latest hot spots on the Web. He or she enjoys the sheer range of possibilities. For such a person, just visiting as many sites as possible and keeping up on the cool ones is an end in itself. The qualitative distinction that staves off levelling for the aesthete is the distinction between those sites that are *interesting* and those that are *boring*, and, thanks to the Net, something interesting is always only a click away. Life consists in fighting off boredom by being a spectator at everything interesting in the universe and in communicating with everyone else so inclined. Such a life produces what we would now call a postmodern self – a self that has no defining content or continuity but is open to all possibilities and to constantly taking on new roles.

But we have still to explain what makes this use of the Web so attractive. Why is there a thrill in being able to be up on everything no matter how trivial? What motivates a passionate commitment to curiosity? Kierkegaard thought that people were addicted to the press, and we can now add the Web, because the anonymous spectator *takes no risks*. The person in the aesthetic sphere keeps open all possibilities and has no fixed identity that could be threatened by disappointment, humiliation, or loss.

Life on the Web is ideally suited to such a mode of existence. On the Internet, commitments are at best virtual commitments. Sherry Turkle has described how the Net is changing the background practices that determine what kinds of selves we can be. In *Life on the Screen*, she details 'the ability of the Internet to change popular understandings of identity'. On the Internet, she tells us, 'we are encouraged to think of ourselves as fluid, emergent, decentralized, multiplicious, flexible, and ever in process'.[29] Thus 'the Internet has become a significant social laboratory for experimenting with the constructions and reconstructions of self that characterize postmodern life'.[30]

Chat rooms lend themselves to the possibility of playing at being many selves, none of whom is recognized as who one truly is, and this possibility is not just theoretical but actually introduces new social practices. Turkle tells us that:

> The rethinking of human . . . identity is not taking place just among philosophers but 'on the ground', through a philosophy in everyday life that is in some measure both proved and carried by the computer presence.[31]

She notes with approval that the Net encourages what she calls 'experimentation' because what one does on the Net has no consequences.[32] For that very reason, the Net frees people to develop new and exciting selves. The person living in the aesthetic sphere of existence would surely agree, but according to Kierkegaard: 'As a result of knowing and being everything possible, one is in contradiction with oneself.'[33] When he is speaking from the point of view of the next higher sphere of existence, Kierkegaard tells us that the self requires not 'variableness and brilliancy' but 'firmness, balance, and steadiness'.[34]

We would therefore expect the aesthetic sphere to reveal that it was ultimately unliveable, and, indeed, Kierkegaard held that, if one leapt into the aesthetic sphere with total commitment expecting it to give one's life meaning, it was bound to break down. Without some way of telling the significant from the insignificant and the relevant from the irrelevant, everything becomes equally interesting and equally boring and one finds oneself back in the indifference of the present age. Writing from the perspective of an aesthete experiencing the despair that signals the breakdown of the aesthetic sphere, he laments: 'My reflection on life altogether lacks meaning. I take it some evil spirit has put a pair of spectacles on my nose, one glass of which magnifies to an enormous degree, while the other reduces to the same degree.'[35]

This inability to distinguish the trivial from the important eventually stops being thrilling and leads to the very boredom the aesthete Net-surfer dedicates his life to avoiding. So, if one throws oneself into it fully, one eventually sees that the aesthetic way of life just doesn't work to overcome levelling. Kierkegaard calls such a realization, despair. Thus, Kierkegaard concludes: 'every aesthetic view of life is despair, and everyone who lives aesthetically is in despair whether he knows it or not. But when one knows it a higher form of existence is an imperative requirement.'[36]

That higher form of existence Kierkegaard calls *the ethical sphere*. In it, one has a stable identity and one engages in involved action. Information is not played with, but is sought and used for serious purposes. As long as information gathering is not an end in itself, whatever reliable information there is on the Web can be a valuable resource serving serious concerns. Such concerns require that people have life plans and take up serious tasks. They then have goals that determine what needs to be done and what information is relevant for doing it.

In so far as the Internet can reveal and support the making and maintaining of commitments for action, it supports life in the ethical sphere. But Kierkegaard would probably hold that the huge number of interest groups on the Net committed to various causes, and the ease of joining such groups, would eventually bring about the breakdown of the ethical sphere. The multiplicity of causes and the ease of making commitments, which should have supported action, will eventually lead either to paralysis or an arbitrary choice as to which commitments to take seriously.

To avoid arbitrary choice, one might, like Judge William, Kierkegaard's pseudonymous author of the description of the ethical sphere in *Either/Or*, turn to facts about one's life to limit one's commitments. Thus, Judge William says that his range of possible relevant commitments is constrained by his abilities, and his social roles as judge and husband. Or, to take a more contemporary example, one could choose which interest groups to join on the basis of certain facts about one's life-situation. After all, there are not merely interest groups devoted to everything from bottle caps to cultural stars like Kierkegaard,[37] there are interest groups, for example, for the parents of children with rare and incurable diseases. So the ethical Net-enthusiast might argue that, to avoid levelling, all one need do is to choose to devote one's life to something that matters based on some accidental condition in one's life.

But the goal of the person in the ethical sphere, as Kierkegaard defines it, is to be morally mature, and Kant held that moral maturity consists in the ability to act lucidly *and freely*. To live ethically, then, one cannot base the meaning of one's life on what accidental facts impose their importance. Thus Judge William is proud of the fact that, as an autonomous agent, he is free to give whatever meaning he chooses to his talents and his roles and all other facts about himself. Thus, he claims that, in the end, his freedom to give his life meaning is not constrained by his talents and social duties, unless he chooses to make them important.

Judge William sees that the choice as to which facts about his life are important is based on a more fundamental choice of what is worthy and not worthy, what is good and what is evil, and that choice is up to him. As Judge William puts it:

> The good *is* for the fact that I will it, and apart from my willing, it has no existence. This is the expression for freedom. . . . By this the distinctive notes of good and evil are by no means belittled or disparaged as merely subjective distinctions. On the contrary, the absolute validity of these distinctions is affirmed.[38]

But Kierkegaard would respond that, if everything were up for choice, including the standards on the basis of which one chooses, there would be no reason for choosing one set of standards rather than another.[39] Besides, if one were totally free, choosing the guidelines for one's life would never make any serious difference, since one could always choose to rescind one's previous choice. A commitment does not get a grip on me if I am always free to revoke it.[40] Indeed, commitments that are freely chosen can and should be revised from minute to minute as new information comes along. The ethical thus breaks down in despair because, either I am stuck with whatever just happens to be imposed on me as important in my life (for example, some life-threatening disease) and so I'm not free, or else the pure power of the freedom to make and unmake commitments undermines itself. As Kierkegaard puts the latter point:

> If the despairing self is *active* . . . it is constantly relating to itself only experimentally, no matter what it undertakes, however great, however amazing and with whatever perseverance. It recognizes no power over itself; therefore in the final instance it lacks seriousness. . . . The self can, at any moment, start quite arbitrarily all over again.[41]

Thus the *choice* of qualitative distinctions that was supposed to support serious action undermines it, and one ends up in what Kierkegaard calls the despair of the ethical. One can take over some accidental fact about one's life and make it one's own only by freely *deciding* that it is crucially important, but then one can equally freely decide it is not, so in the ethical sphere all meaningful differences are levelled by one's making one's freedom absolute.

According to Kierkegaard, one can only stop the levelling of commitments by being *given* an individual identity that opens up an individual world. Fortunately, the ethical view of commitments as freely entered into and always open to being revoked does not seem to hold for those commitments that are most important to us. These special commitments are experienced as gripping our whole being. Political and religious movements can grip us in this way, as can love relationships and, for certain people, such 'vocations' as science or art. When we respond to such a summons with what Kierkegaard calls infinite passion – that is, when we respond by accepting an *unconditional commitment* – this commitment determines what will be the significant issue for us for the rest of our life. Such an unconditional commitment thus blocks levelling by establishing qualitative distinctions between what is important and trivial, relevant and irrelevant, serious and playful in my life. Living by such an irrevocable commitment puts one in what Kierkegaard called the *Christian/religious sphere of existence.*[42]

But, of course, such a commitment makes one vulnerable. One's cause may fail. One's lover may leave. The detached reflection of the present age, the hyperflexibility of the aesthetic sphere, and the unbounded freedom of the ethical sphere are all ways of avoiding one's vulnerability, but it turns out, Kierkegaard claims, that, for that very reason, they level all qualitative distinctions, and end in the despair of meaninglessness. Only a risky unconditional commitment and the strong identity it produces can give an individual a world organized by that individual's unique qualitative distinctions.

This leads to the perplexing question: what role if any can the Internet play in encouraging and supporting unconditional commitments? A first suggestion might be that the movement from stage to stage would be facilitated by living experimentally on the Web, just as flight simulators help one learn to fly. One would be solicited to throw oneself into enjoying Net surfing until one found that boring, then into freely choosing which interest group was important until that choice revealed its absurdity, and so finally one would be driven to let oneself be drawn into a risky unconditional commitment as the only way out of despair. Indeed, at any stage, from looking for all sorts of interesting Websites as one surfs the Net, to striking up a conversation in a chat room, to joining an interest group to deal with an important problem in one's life, one might just find oneself being drawn into a lifetime commitment. No doubt this might happen – people who meet in chat rooms may fall in love – but it is highly unlikely.

Kierkegaard would surely argue that, while the Internet, like the public sphere and the press, does not *prohibit* unconditional commitments, in the end, it *undermines* them. Like a simulator, the Net manages to capture everything but the risk.[43] Our imaginations can be drawn in, as they are in playing games and watching movies, and no doubt, if we are sufficiently involved to feel we are taking risks, such simulations can help us acquire skills, but in so far as games work by temporarily capturing our imaginations in limited domains, they cannot simulate serious commitments in the real world. Imagined commitments hold us only when our imaginations are captivated by the simulations before our ears and eyes. And that is what computer games and the Net offer us. But the risks are only imaginary and have no long-term consequences.[44] The temptation is to live in a world of stimulating images and simulated commitments and thus to lead a simulated life. As Kierkegaard says of the present age, 'it transforms the task itself into an unreal feat of artifice, and reality into a theatre'.[45]

The test as to whether one had acquired an unconditional commitment would come only if one had the passion and courage to transfer what one had learned on the Net to the real world. Then one would confront what Kierkegaard calls 'the danger and life's stern judgment'. But precisely the attraction of the Net, like that of the press in Kierkegaard's time, is that it inhibits that final plunge. Indeed, anyone using the Net who was led to risk his or her real identity in the real world would have to act against the grain of what attracted him or her to the Net in the first place.

So it looks like Kierkegaard may be right. The press and the

Internet are the ultimate enemy of unconditional commitment, but only the unconditional commitment of what Kierkegaard calls the religious sphere of existence can save us from the nihilistic levelling launched by the Enlightenment, promoted by the press and the public sphere, and perfected in the World Wide Web.

Conclusion

We have now seen that our body, including our emotions, plays a crucial role in our being able to make sense of things so as to see what is relevant, our ability to let things matter to us and so to acquire skills, our sense of the reality of things, our trust in other people, and finally, our capacity for making the unconditional commitments that give meaning to our lives. It would be a serious mistake to think we could do without these embodied capacities – to rejoice that the World Wide Web offers us the chance to become more and more disembodied, detached ubiquitous minds leaving our situated, vulnerable bodies behind.

We've seen that there is always a trade-off between what the Web offers and what it takes away. In the case of the hyperlinks discussed in Chapter 1, the trade-off is what one might call symmetrical or reciprocal. There is something to be said both for relevance and for scope, for quality and for quantity of information. Neither is intrinsically superior, but more of one means less of the other. So one must simply choose which one prefers, depending on one's practical needs and whether one's sensibilities are modern or postmodern.

In the other three chapters, however, the trade-offs are more complicated. The two options are not equal; one side of the trade-off is superior to the other. One might call these asymmetrical trade-offs.

In Chapter 2 we saw that, as far as education is concerned, the Net can be useful in supplying the facts and rules as well as the drill and practice required by a beginner. It seems, however, that the involvement and risk that come from making interpretations that can be mistaken and learning from one's mistakes are necessary if one is to acquire expertise. Such involvement is absent if one is just sitting alone in front of one's computer screen looking at a lecture downloaded from the Web. There is more involvement in an on-line *interactive* lecture on the Web, but the sense of taking a risk and accepting approval or criticism in front of others is much reduced, and, therefore, so is the involvement. Such lectures are, therefore, not likely to produce more than competence. Only in a classroom where the teacher and learner sense that they are taking risks in each other's presence, and each can count on criticism from the other, are the conditions present that promote acquiring proficiency, and only by acting in the real world can one acquire expertise. As for the apprenticeship necessary to becoming a master, it is only possible where the learner sees the day-to-day responses of a master and learns to imitate her style.

Thus, we saw in Chapter 2 that, in considering distance education, one has to choose between economy and efficacy, and that, while administrators and legislators tend to prefer the 'maximum throughput' even if it can only produce competence, most teachers, parents, and students, if they can afford it, would prefer the shared involvement that produces proficiency, and the real-world experience and mentoring that makes possible the acquisition of expertise and mastery.

Where our sense of reality is concerned, the trade-off is differently asymmetrical. The relation of presence to telepresence is not a question of the advantages and disadvantages of each, and so of choosing one over the other. Rather, telepresence presupposes presence. Here, the asymmetry is one of dependence. Thus, I argued that telepresence, both of objects and people, is parasitical on a robust sense of the presence of the real correlative with the body's set to cope with things and people. We may lament the risks endemic to an embodied world where we are embedded with objects and others in local situations, but the idea of living in boundless Cyberia, where everyone is telepresent to everyone and everything, makes no sense.

Finally, where meaning is concerned, again the trade-off is asymmetrical. This time, one side is positive and the other negative. If we remain the kind of beings that Kierkegaard understood us to be, we will despair if all meaningful distinctions are levelled, and since meaningful distinctions require commitment and vulnerability, which require our embodied finitude, we should have no trouble in choosing between disembodied nihilism and embodied meaning.

But isn't all this just to say that we can see what the Web can't do for us, but there may be great things it *can* do that we can't yet even imagine. After all, in the 'Phaedo' Plato famously objected to the introduction of writing as opposed to speech, because, as he pointed out, writing reduces the richness of communication since it makes it impossible to read the speaker's tone and bodily posture. Furthermore, he saw that, if agreements could be made at a distance, they would not be as binding as agreements sealed by the spoken word. He also thought that people would lose their ability to remember important events.

Of course, all of that was true, but Plato couldn't foresee that, thanks to writing, we would gain a wider range of communication, new ways of making contracts at a distance, and a whole new cultural memory. If he could have foreseen all this, he might well have had a more positive view of the trade-offs involved.

No doubt the Internet, like the car, will have huge consequences both for good and ill that we cannot foresee. None the less, there are two important differences between my argument and Plato's. I don't know what the claims for the value of writing among Plato's contemporaries were, but I have been arguing that the positive claims for the value of the Internet offered by our contemporaries are mostly hype. Whatever the long-range value of the Net turns out to be, it won't be the quality of information it offers, the democratic distance learning it makes possible, the presence of the Net user to all of reality, and the possibility of a new life full of meaning.

More importantly, if my arguments are right, the Net differs dramatically from writing as to what an uncritical use of it could lead us to lose. It's unlikely that any of Plato's contemporaries were proposing that everyone would be better off the more they gave up talking and lived their lives through writing, whereas we are being told by groups like the Extropians that, the more we can give up our bodies and live in cyberspace, the better off we will be. My answer is that, if we managed to live our lives in cyberspace, we would lose a lot more than the face-to-face conversations, verbal promises, and memory power Plato saw were endangered by writing. We would lose our only reliable way of finding relevant information, the capacity for skill acquisition, a sense of reality, and the possibility of leading meaningful lives – the last three of which are constitutive of us as human beings. Indeed, they are so definitive of who we are that nothing new and unexpected could possibly make up for our losing them.

But we would, of course, still like to know what the Web is good for and what it is not, so we can use it for what it does well. How then can we profit from the Web in each of the above areas? Obviously, we need to foster a symbiosis in which we use our bodies and their positive powers, to find what is relevant, learn skills through involvement, get a grip on reality, and make the risky commitments that give life meaning, while letting the Web contribute its amazing capacity to store and access astronomical amounts of information, to connect us to others, to enable us to be observers of far-away places, and to experiment without risk with other worlds and selves. In place of a summary of what has already been said, then, I'd like to offer a few examples of how this symbiosis might work.

1 RELEVANCE AND RETRIEVAL

The failure of AI and the poor performance of syntactic systems for information retrieval have shown that, if we managed to give up our embodiment for a limitless life in cyberspace, we would also have to give up being able to retrieve most of the information we needed. Since Web crawlers and search engines respond only to syntax and not meaning, they can at best return only a small percentage of the potentially relevant web pages to the user. A dramatic way to see this is to consider the ongoing war between the search-engine designers and the Spammers, those who want their irrelevant information to replace the relevant information the Net user is seeking.

Website designers would like to make their sites the most likely to be picked up by search engines looking for certain kinds of content. There are, in fact, services that tell people how to bias their sites so that they will be picked up by popular search engines. Such services propose a variety of techniques, the most widely publicized of which is 'keyword stuffing', where pages and pages of a single word are added invisibly to the Website to bias the search engines that count word frequencies. In self-defence, search engines are now programmed to ignore author-selected key words, although that is one of the most successful syntactic ways of ascertaining the content of a document or site. In response, the Spammers now embed the key words in computer-generated strings of words that have the grammatical form of meaningful sentences. A famous example from linguistics of such a sentence is 'Colourless green ideas sleep furiously.' The designers are thus trying to influence the search engines, and the search engines are trying not to be influenced by the designers. This is a fascinating war that should be the subject of a whole book, but the philosophical moral is simple: you can fool a search engine because it is a syntax machine and so has no common sense and no way to pick up on meaning, but you can't similarly fool a human being.

A human being responding to the meaning of a document or Website would not be fooled by pages and pages of the same word, or by meaningless grammatical sentences, while a search engine that can only use formal measures to find content inevitably falls for such tricks. This suggests that human beings should play a role in search. Gordon Rios predicts that a new group of professional intermediaries will soon arise who respond to meaning and have common sense and so are able to spot the tricks of the Spammers. Human beings could also take advantage of their knowledge of how various search engines work so as to use whichever one is most appropriate for each specific request. Most important of all, these intermediaries would be experts in some specific domain and, like the librarians of old, would be able to select the relevant documents from among all those the search engine returns. An article in *The New York Times* sums up the realization that embodied people with their sense of relevance cannot be dispensed with but need to form a symbiotic relation with the disembodied machines:

> When search engines first appeared, they were hailed for accomplishing two things that could not be done by people on any large scale: Search engines used software agents to find and index sites almost as soon as they appeared. And they could almost instantaneously match a far-flung Web page with a single keyword typed into a beckoning search box.
>
> But the promise of automation has been tempered by the Web's success. There are now more than one billion Web pages, and according to some experts' calculations, the number has been doubling once every eight months. . . . To cope, many search engines have concluded that simply indexing more pages is not the answer. Instead, they have decided to rely on the one resource that was once considered a cop-out: human judgment.[1]

2 DISTANCE LEARNING

Granted that acquiring skills requires involvement and risk, and that professional and cultural skills can be passed on only from body to body by means of apprenticeship, still, in education, there are many ways of combining the advantages of old-fashioned lecture/discussions with the power of the Web. Given a class in which students are bodily present and there is already a shared mood of concern for learning, teachers have found that putting their assignments, questions, paper topics, etc., on a course Website helps students stay informed as to what is going on in the course. Teachers can also pose questions that the students can discuss in a news group, and they can intervene in the discussion when necessary to clarify issues raised by the students.

In addition, I've found that it's useful to put my actual lectures in MP3 format on my course Website, so that students who have to miss class can listen later from their dorms, and students writing papers can review lectures that went by too quickly for them to follow. Of course, if one discusses films in one's courses, as I do, one can also include on the course Website film clips of the scene being discussed cued to the audio.

This semester I've gone a step further and arranged to webcast one of my courses in video so that students can watch the course from their dorms rather than sit on the floor in a crowded classroom. One might wonder why, in such a case, students would bother to come to class at all, but most students must be getting something special out of being bodily present at the lectures – sharing the mood in the room and making risky suggestions in class discussion – since, although they can now watch the lectures on their computers any time that is convenient, class attendance has barely been affected, except on rainy days, when attendance drops by about 30 per cent. This suggests that presence in class is felt to be such a positive experience that most students will slog through bad weather to attend, but that the Webcast offers enough so that those who cherish their comfort can make do with distance learning – it is certainly better than nothing.

Students who watched my lectures from their dorms said that they found the archived Webcasts helpful once they had been *to the lecture*. Thereafter, they could replay the Webcast, stopping the video to go over difficult points. But they felt that there was something about being present in the room with the lecturer and the other students that give them a sense of 'interconnectedness' that they would not want to do without. They also felt that the presence of the lecturer focused them on what was important in the material being presented. Yet they also said, to my surprise, that they preferred the *audio* version of my lectures. They said they found the moving image merely distracting. They appreciated seeing the charts and outlines, but they didn't need to see me writing them on the board since I always make the material available on the course Website. Of course, I put more on the board than just charts, so, in response to the students' comments, I'm considering taking a digital picture of the board at the end of each lecture and putting that picture on the course Website along with the audio. Then, it seems, there would be no added value gained in Webcasting my course. In any case, for these students, distance learning seems to presuppose bodily presence.

3 TELEPRESENCE

I've argued that telepresence can never give us a sense of the reality of far-away things, nor can it convey a sense of trust of distant human beings. It therefore seems a waste of effort to try and make telepresence do the job of bodily presence by adding feeling, smells, etc. Still, as we have seen, there is a place for teleconferencing when people already know and trust each other. And, of course, telepresence is still indispensable in those areas for which it was developed, such as dealing with things where bodily presence is too big, too small, too risky, etc., as in repairing nuclear reactors and exploring unliveable planets. These possibilities predate the World Wide Web, but the Web can expand our perceptions and active intervention to the far corners of the universe. It is estimated that there are now over 2,000 Webcams in operation, and, through them, one can see the traffic or the weather at any time almost anywhere in the world. Our minds can, indeed, expand to more and more of the universe. As long as we continue to appreciate our bodies and don't lose our engineering expertise by substituting distance learning for lectures and apprenticeship, we can look forward to improved versions of vehicles like the Mars rover that will explore distant planets with millions of us televiewers on board.

But we have to be wary of the attraction of telepresence. The use of robots along with Internet-mediated telepresence offers the attractive possibility of each of us being able to control distant representatives of ourselves that are extensions of our eyes and ears. We could then take part in situations too dangerous for us to explore in person, by, for example, walking into a nuclear reactor, or we could simply be present in situations that we were too far away from to attend, such as taking part in the Oscar awards ceremony while shooting on a set in France. After forty years of being told that a household robot is just around the corner, one would think that such a robot slave who could represent us in dangerous and far-away places would be easy to build, and now, thanks to the possibility of telepresence, easy to control. After all, thanks to Ken Goldberg's Telegarden, one already can control a robot arm so as to plant and water a seed in Linz.

Sadly, however, reality lags far behind predictions. At an international meeting of robot makers that recently took place at MIT, all but a few fanatics agreed that humanoid robots would, for a long time, remain science fiction. A *New York Times* reporter filed the following report:

> Last month's organizers of the *Humanoids 2000* conference surveyed some of the participants about possible social implications of their work. On a scale of 0, for highly unlikely, to 5, for highly likely, the robotics researchers rated the possibility that robots 'will be the next step in evolution and will eventually displace human beings' a zero. 'They are much less euphoric than other people, say, movie producers', said Dr. Alois Knoll . . . one of the organizers of the conference. . . . Dr. Knoll listed the limitations of present-day robots: 'We don't have the mechanical dexterity. We don't have the power supply. We don't have the brains. We don't have the emotions. We don't have the autonomy in general . . . to even come close to humans.'[2]

But not to worry, Ken Goldberg and his co-workers have suggested a solution that is now being explored by the MIT Media Lab. Those involved realize that robots will, for a long time, be too clumsy to be our representatives, so they propose that we recruit real people to do the job. They are therefore working on how a person could teleguide a Tele-Actor wearing the webcams and microphones that would enable the controller to be telepresent at far-away events. The Tele-Actor, impersonating a robotic slave, would wear goggles with lights around the edges that would signal to him or her which way to turn and how fast to move, etc. as the controller teleguided 'it' to take part, for example, in a far-away award ceremony.

Fortune Magazine published the following report on the Media Lab project under the title 'Being-There'.

> Send a Tele-Actor out to a location, and you see what it sees and hear what it hears. Multiple participants can log on, all sharing the same viewpoint, all helping to direct the action. 'It lets anyone tap into a remote experience – a sports event, a conference, maybe even a place too dangerous for most people, like a war zone', says Ken Goldberg. . . . Goldberg created the idea with a team of colleagues as part of their experiments in 'telepresence', which uses technology to break down distance. As bandwidth improves and camera tech gets cheaper, they see Tele-Actors becoming common.[3]

By proposing an ingenious end-run around the failures of AI and the setbacks of humanoid robot research, the Media Lab has succeeded in once again illustrating a disturbing tendency of computer enthusiasts. Computers exhibit the possibility of augmenting human capacities such as memory and calculating ability, but it turns out they lack other abilities such as intelligence and the ability to move their body in adaptive and coordinated ways. So, since computers can't do what people can easily do, to get the desired collaboration between person and machine, people will have to be trained to act like computers. To take advantage of the possibility of telepresence, since robots can't be programmed to behave like people, people will have to learn to behave like robots.

4 MEANING

Some of the most vexing questions arise over whether the World Wide Web is improving or diminishing the quality of our lives. We've seen that two surveys suggest that living through the Net leads to isolation, and one of these surveys finds, in addition, that use of the Net leads to loneliness and depression.

Yet a recent National Public Radio survey showed that people felt just the opposite of the ill-effects found in the Carnegie-Mellon and Stanford studies. I quote from the NPR Website:

> A new poll by National Public Radio, the Kaiser Family Foundation, and Harvard's Kennedy School of Government shows that people overwhelmingly think that computers and the Internet have made Americans' lives better. Nearly 9 in 10 say computers have made life better for Americans, and more than 7 in 10 say the Internet has made life better.[4]

Yet the poll also showed that 'more than half [of those polled] say computers have led people to spend less time with their families and friends'. This shows, I think, that not only are we transformed by the way we use our tools; we are not aware of how we are being transformed, so we need all the more to try to make explicit what the Net is doing for us and what it is doing to us in the process.

I've suggested that, where meaning is concerned, what the Net is doing to us is, in fact, making our lives worse rather than better. Living one's life on the Web is attractive because it eliminates vulnerability and commitment but, if Kierkegaard is right, this lack of passion necessarily eliminates meaning as well.

It should thus be clear that tools are not neutral, and that using the Net diminishes one's involvement in the physical and social world. This, in turn, diminishes one's sense of reality and of the meaning in one's life. Indeed, it seems that, the more we use the Net, the more it will tend to draw us into the unreal, lonely, and meaningless world of those who want to flee all the ills that flesh is heir to.

If, however, one is already committed to a cause, the World Wide Web can increase one's power to act, both by providing relevant information, and by putting committed people in touch with other people who share their cause and who are ready to risk their time and money, and perhaps even their lives, in pursuing their shared end. The landmine treaty, for example, was hammered out and promoted largely thanks to the fact that the Web is international and has no gatekeepers.

But, the risk posed by the ambiguous similarity of social cyberspaces to communities in the embodied social world comes out clearly in the second edition of Howard Rheingold's influential book, *The Virtual Community*.[5] In his new chapter, 'Rethinking Communities', Rheingold responsibly discusses a tangle of issues surrounding the advantages and disadvantages of many–one interactions in cyberspace. Unfortunately, his analysis is marred by his failure to distinguish the various forms such Internet communities can take.

To begin with, Rheingold defends his conviction that cybercommunities could improve democracy. 'The most serious critique of this book', he says, 'is the challenge to my claim that many–one-discussions could contribute to the health of democracy by making possible better communications among citizens.'[6] He then goes on to develop the claim made in the first edition that the Net 'might help revitalize the public sphere', indeed, that 'the vision of a citizen-designed, citizen-controlled worldwide communications Network is a version of technological utopianism that could be called the vision of "the electronic agora"'. 'In the original democracy, Athens', he explains, 'the agora was the market-place, and more – it was where citizens met to talk, gossip, argue, size each other up, find the weak spots in political ideas by debating about them.'[7]

But the vision of a *worldwide* electronic agora precisely misses the Kierkegaardian point that the people talking to each other in the Athenian agora were members of a direct democracy who were directly affected by the issues they were discussing, and, most importantly, the point of the discussion was for them to *take the responsibility and risk of voting publicly* on the questions they were debating. For Kierkegaard, a world-wide electronic agora is an oxymoron. The Athenian agora is precisely the opposite of the public sphere, where anonymous electronic kibitzers from all over the world, who risk nothing, come together to announce and defend their opinions. As an extension to the deracinated public sphere, the electronic agora is a grave danger to real political community. Kierkegaard enables us to see that the problem is not that Rheingold's 'electronic agora' is too utopian; it is not an agora at all, but a nowhere place for anonymous nowhere people. As such, it is dangerously distopian.

The discussion is blurred by the fact that Rheingold does not distinguish the negative influence of the contribution of the Net to the public sphere from two positive ways in which the Net allows people to leap out of the prison of endless reflection: the *aesthetic possibilities* of virtual commitments, on the one hand, and the *ethical actuality* of committed action, on the other.

Virtual communities constitute an interesting leap into the aesthetic sphere of existence. Such communities are in a certain way the antithesis of the public sphere since passionate commitments are encouraged, not frowned upon, and the issues debated are of crucial concern to the virtual community. Kierkegaard agrees that people in the aesthetic sphere of existence are involved in each other's emotional lives. But what is essential to him is that, although the aesthetic person lives in a world of intense feeling and lively communication, all the drama is like a game in that it has no real-world consequences and there is no real-world risk. Individuals can enter or leave a virtual community much more easily than they can move out of a town they dislike. As we saw, Kierkegaard says that the aesthetic sphere turns existence into a play.

Rheingold frankly faces the danger 'that virtual communities might be bogus substitutes for true civic engagement'.[8] And he acknowledges that:

> most of what needs to be done has to be done face to face, person to person – civic engagement means dealing with your neighbors in the world where your body lives. . . . Discourse among informed citizens can be improved, revived, restored to some degree of influence – but only if a sufficient number of people learn how to use communication tools properly, and apply them to real-world political problem-solving.[9]

One could conclude, and Rheingold might well agree, that, as a game, involvement in virtual communities is not a threat to political engagement in one's actual community. But it becomes harmful if, as is often the case, its risk-free nature makes it more attractive than the dangerous real world, and so drains off the time and energy that citizens could have given to actual community concerns.

So, in his new chapter, Rheingold's emphasis shifts to the role the Internet can play in bringing together people with concrete problems and enabling them to act more effectively.

Thus, he proposed 'experimenting with different tools for civic involvement'.[10] But his defence of such Internet interest groups is presented as a defence of the public sphere, so that the important distinction between detached and anonymous talk and involved responsible action is lost. Rheingold's impressive list of Internet groups that foster concrete commitments – such as a group called Cap-Advantage that provides 'Tools for Online Grassroots Advocacy and Mobilization' – also includes free-floating public sphere groups like Freedom Forum, which he describes as 'a nonpartisan international foundation dedicated to free press, free speech and free spirit for all people'.[11]

If in reading Rheingold's book one bears in mind Kierkegaard's threefold distinction between the public sphere with its reflective detachment from local issues, the aesthetic sphere with its risk-free simulation of the serious concerns of the real world, and the ethical sphere with its local political commitments, one can then be grateful to Rheingold for laying out the impressive spectrum of what the Net can provide. But one needs to bear in mind, besides the above Kierkegaardian categories, Kierkegaard's account of the religious sphere of unconditional commitment, before attempting to pose, let alone resolve, the serious social issues the Net raises.

In sum, as long as we continue to affirm our bodies, the Net can be useful to us in spite of its tendency to offer the worst of a series of asymmetric trade-offs: economy over efficiency in education, the virtual over the real in our relation to things and people, and anonymity over commitment in our lives. But, in using it, we have to remember that our culture has already fallen twice for the Platonic/Christian temptation to try to get rid of our vulnerable bodies, and has ended in nihilism. This time around, we must resist this temptation and affirm our bodies, not in spite of their finitude and vulnerability, but because, without our bodies, as Nietzsche saw, we would be literally nothing. As Nietzsche has Zarathustra say: 'I want to speak to the despisers of the body. I would not have them learn and teach differently, but merely say farewell to their own bodies – and thus become silent.'[12]

Notes

INTRODUCTION

1 The Extropians are very far out, but the same ideas show up in supposedly serious books such as Hans Moravec's *Mind Children*, Cambridge, MA, Harvard University Press, 1988.

2 A tendency that Martin Heidegger claims is definitive of our modern understanding of what it is to be anything at all. See Martin Heidegger, 'The Question Concerning Technology', in *The Question Concerning Technology*, New York, Harper and Row, 1977.

3 Not that there haven't been real innovations. New ways of linking information have transformed libraries; course Websites in colleges and universities have made it possible for students to hear lectures and engage in discussions without leaving their rooms; telerobotics has made it possible to control a vehicle on Mars and one day millions of spectators will no doubt be able to look out of such a vehicle as it moves across the Mars surface; and e-mail has opened up surprising new possibilities, from political dissidents working together for reform, to proud grandparents sending their friends the latest digital photos of their grandchildren. But all these surprising new developments are minor compared to what has been predicted.

4 A. Harmon, 'Researchers Find Sad, Lonely World in Cyberspace', *The New York Times*, August 30, 1998. Harmon continues:

> Those participants who were lonelier and more depressed at the start of the two-year study, as determined by a standard questionnaire administered to all subjects, were not more

likely to use the Internet. Instead, Internet use itself appeared to cause a decline in psychological well being, the researchers said.

5 R. Kraut, M. Patterson, V. Lundmark, S. Kiesler, T. Mukophadhyay and W. Scherlis, 'Internet Paradox: A Social Technology that Reduces Social Involvement and Psychological Well-being?' *American Psychologist*, 1998, vol. 53, no. 9, pp. 1017–31.

6 Ibid. It seems that lack of physical presence can lead to a kind of moral isolation too. When Larry Froistad confessed to his e-mail support group that he had murdered his daughter, the members of the group offered him sympathy; only one felt they should turn him over to the police. See, 'On-Line Thoughts on Off-Line Killing' by Amy Harmon, *The New York Times*, April 30, 1998. 'It seemed to Ms. De Carlo that the nature of on-line communication – which creates a psychological as well as physical distance between participants – was causing her friends to forget their off-line responsibilities to bring a confessed murderer to justice.'

7 J. P. Barlow, 'A Declaration of the Independence of Cyberspace', Davos, Switzerland, February 8, 1996, http://members.iquest.net/˜dmasson/barlow/Declaration-Final.html

8 Moravec, op. cit.

9 R. Kurzweil, *The Age of Spiritual Machines*, New York, Penguin, 2000.

10 E. Dyson *et al.*, 'Cyberspace and the American Dream: A Magna Carta for the Knowledge Age. Release 1.2', Washington, DC, The Peace and Progress Foundation, 1994.

11 Plato, 'Gorgias', 492e7–493a5. Socrates says: 'I once heard one of our wise men say that we are now dead, and that our body (*soma*) is a tomb (*sema*).'

12 Plato, 'Phaedo', *The Last Days of Socrates*, Baltimore, MD, Penguin, 1954, p. 84.

13 F. Nietzsche, *Thus Spake Zarathustra*, trans. W. Kaufmann, New York, Viking Press, 1966, p. 35.

14 Ibid., p. 34.

ONE THE HYPE ABOUT HYPERLINKS

1 National Public Radio, 'The Future of Computing', *Talk of the Nation, Science Friday*, July 7, 2000.

2 S. Lawrence and C. L. Giles, NEC Research Institute, 'Searching the World Wide Web', *Science*, 280, April 3, 1998, p. 98. Moreover, the size isn't just the number of Websites or pages; the number of hyperlinks embedded in the Web pages is even larger.

3 There has been some interesting litigation of late trying to stop this 'free-linking' of anything to anything, in which parties have sued others who made links to the plaintiff's Web page. Of course, this is a fraction of a fraction of a per cent, and is unlikely to have any significant effect on the way the Web is run which has been called a 'loose ad-hocracy'. It no doubt just reflects the dying gasp of the old guard who would like to place at least *some* limits on the eventual linking of everything to everything.

4 The Dewey decimal system was organized in this way. It did not even allow the same item to be filed under two different categories, but now librarians have more leeway and file the same information under several different headings. For example, Philosophy of Religion would presumably be filed under Philosophy and Religion. Still, however, there is an agreed-upon hierarchical taxonomy.

5 David Blair's book, *Language and Representation in Information Retrieval*, New York, Elsevier Science, 1990, was chosen 'Best Information Science Book of the Year' in 1999 by the American Society for Information Science, and Blair himself was named 'Outstanding Researcher of the Year' by the same society in the same year.

6 Blair adds that, 'while we think of organizing information for a particular use or practice, we forget that the use or practice helps to organize the information, too. I call these "Natural Sets" of information. So the selection and organization of information to serve a practice is an interactive process. What the Web prevents or discourages is this sort of natural interplay between information and practice.'

7 What people now refer to as the modern subject came into being in the early seventeenth century as – thanks to Luther, the printing press, and the new science – people began to

think of themselves as self-sufficient individuals. Descartes introduced the idea of the subject as what underlay changing mental states, and Kant argued that, as the objectifier of everything, the subject must be free and autonomous. As we shall see in Chapter 4, Søren Kierkegaard concluded that each one of us is a subject called upon to take on a fixed identity that defines who one is and what is meaningful in one's world. (See Chapter 4.)

8 *The New York Times*, January 9, 2000.

9 Draft of David Blair's forthcoming book, *Wittgenstein, Language and Information*.

10 D. Blair, 'Will it Scale up? Thoughts about Intellectual Access in the Electronic Networks', in A. Okerson (ed.), *Gateways, Gatekeepers, and Roles in the Information Omniverse*, Washington, DC, Association of Research Libraries: Office of Scientific and Academic Publishing, 1994.

11 See H. Dreyfus, *What Computers (Still) Can't Do*, 3rd edn, Cambdrige, MA, MIT Press, 1992.

12 See D. Lenat and R .V. Guha, *Building Large Knowledge-Based Systems*, New York, Addison Wesley, 1990.

13 Ibid., p. 4.

14 V. Pratt, *CYC Report*, Stanford University, April 16, 1994.

15 R. V. Guha and W. Pratt, 'Microtheories: An Ontological Engineer's Guide', MCC Technical Report Number CYC-050–92, 1992, p. 15.

16 Ibid.

17 R. V. Guha and A. Y. Levy, 'A Relevance Based Meta Level', MCC Technical Report Number CYC-040–90, 1990, p. 7.

18 Ibid.

19 In a conversation, Guha told me that he and Lenat found that they needed hundreds of relevance axioms and, after formulating a thousand or so, they gave up. The task thus looks hopeless since, if one needs so many relevance axioms, one would eventually need higher-order relevance axioms to determine, in any given case, which relevance axioms were relevant.

20 D. Swanson (former Dean of the Library School at the University of Chicago), 'Historical Note: Information Retrieval and the

Future of an Illusion', *Journal of the American Society for Information Science*, vol. 32, no. 2, 1998, pp. 92–8.

21 The role of the body in our being able to experience space, time and objects is worked out in detail in S. Todes, *Body and World*, Cambridge, MA, MIT Press, 2001.

22 Personal communication.

23 G. Rios, personal communication. My italics. Rios adds: 'Some of the engines boost performance by using proximity search which scores a document higher if the search words are found closer together in a document. But, again, what is the proper way to score proximity? It varies by the context and subject, of course, in a way that may be transparent to a person, but is opaque to a computer.' He notes further that many queries (about half) only get issued one time, and for those the performance is probably about half of the above.

24 D. Swanson, op. cit.

TWO HOW FAR IS DISTANCE LEARNING FROM EDUCATION?

1 T. Oppenheimer, 'The Computer Delusion', *The Atlantic Monthly*, July 1997.

2 See Dreyfus and Dreyfus, *Mind over Machine*, New York, Free Press, 1988, Chapter 5.

3 The most surprising endorsement comes from former United States Secretary of Education, William Bennett. As reported in the *New York Times* of December 28, 2000, Bennett once said, 'When you hear the next pitch about cyber-enriching your child's education, keep one thing in mind: so far, there is no good evidence that most uses of computers significantly improve learning.' But recently 'Mr. Bennett announced that he was founding a for-profit school called K12 that would offer a child a complete elementary and secondary education, from kindergarten through 12th grade, all of it online.' As if confirming the fears of those who think that children educated by computers will lose touch with the texture of the real world, 'his school's kindergartners will use mouse pads to finger paint and older students will do virtual chemistry experiments with

animated beakers and electronic Bunsen burners.' It seems that this optimism is shared in China. Reuters reports on August 22, 2000: 'Chinese President Jiang Zemin offered a ringing endorsement of the Internet on Monday, saying e-mail, e-commerce, distance learning and medicine would transform China.'

4 This speech was given on April 29, 1999, at the Networking '99 conference in Washington, DC. It was published in *Educom Review*, vol. 34, no. 6, 1999, and can be found on the Web at: http://www.educause.edu/ir/library/html/erm9963.html

5 Here are Hundt's remarks on this last point: 'The Internet is also an assault on elites. One of the top three Ivy Schools reported the other day it was a little worried about the skewing of students with respect to the income base. . . . Here was the statistic. Eighty-five percent of the students at this top Ivy League school . . . were from upper-income families or higher. . . . It's high time for the highest level of education to be democratized in this country'.

6 'The Paula Gordon Show', broadcast on February 19, 2000, on WGUN.

7 T. Gabriel, 'Computers Can Unify Campuses, But Also Drive Students Apart', *The New York Times*, November 11, 1996.

8 For more details, see Dreyfus and Dreyfus, op. cit.

9 See M. Polanyi, *Personal Knowledge*, London, Routledge & Kegan Paul, 1958.

10 Patricia Benner has described this phenomenon in *From Novice to Expert: Excellence and Power in Clinical Nursing Practice*, Menlo Park, CA, Addison-Wesley, 1984, p. 164. Furthermore, failure to take risks leads to rigidity rather than the flexibility we associate with expertise. When a risk-averse person makes an inappropriate decision and consequently finds himself in trouble, he tries to characterize his mistake by describing a certain class of dangerous situation and then makes a rule to avoid that type of situation in the future. To take an extreme example, if a driver, hastily pulling out of a parking space, is side-swiped by an oncoming car he mistakenly took to be approaching too slowly to be a danger, he may resolve to follow the rule, never pull out if

there is a car approaching. Such a rigid response will make for safe driving in a certain class of cases, but it will block further skill refinement. In this case, it will prevent acquiring the skill of flexibly pulling out of parking places. In general, if one seeks to follow general rules one will not get beyond competence. Progress is only possible if, responding quite differently, the driver accepts the deeply-felt consequences of his action without detachedly asking himself what went wrong and why. If he does this, he is less likely to pull out too quickly in the future, but he has a much better chance of ultimately becoming, with enough frightening or, preferably, rewarding experiences, a flexible, skilled driver.

One might object that this account has the role of emotions reversed; that the more the beginner is emotionally committed to learning, the better, while an expert could be, and, indeed, often should be, coldly detached and rational in his practice. This is no doubt true, but the beginner's job is to follow the rules and gain experience, and it is merely a question of motivation whether he is involved or not. Furthermore, the novice is not emotionally involved in *choosing* an action, even if he is involved in its outcome. Only at the level of competence is there an emotional investment in the *choice of action*. Then emotional involvement seems to play an essential role in switching over from what one might roughly think of as a left-hemisphere analytic approach to a right-hemisphere holistic one. Of course, not just any emotional reaction, such as enthusiasm, or fear of making a fool of oneself, or the exultation of victory, will do. What matters is taking responsibility for one's successful and unsuccessful choices, even brooding over them; not just feeling good or bad about winning or losing, but replaying one's performance in one's mind step by step or move by move. The point, however, is not to *analyse* one's mistakes and insights, but just to *let them sink in*. Experience shows that only then will one become an expert. After one becomes an expert one can rest on one's laurels and stop this kind of obsessing, but if one is to be the kind of expert who goes on learning, one has to go on dwelling emotionally on what critical choices one has made and how they affected the outcome.

11 K. Nielsen, 'Musical Apprenticeship, Learning at the Academy of Music as Socially Situated', *Nordic Journal of Educational Research*, vol. 3, 1997.

12 If we take a closer look at apprenticeship, we find that this kind of training contains important insights for testing as well as teaching. The apprentice becomes a master by imitating a master. He gradually learns how to do the whole task. Since skills are not learned by components but, rather, by small holistic improvements, there is no way to test the student in each component of the relevant skill. Where mastery is at stake, the kind of examination used in most universities and necessarily on the Internet is not useful, and even counter-productive. Rather, instead of giving the apprentice periodic examinations to see if he has mastered the components that are normally mastered by students at his stage, when it seems to the master that an apprentice has learned his craft, he is asked to do what is normally done by an expert in his domain of expertise. For example, if he is learning to make a musical instrument, he may be asked to make, say, a violin. But, without an examination scored on a normal curve, who is to decide whether or not the apprentice has made a good violin? Only an expert can tell. So the masters gather around and play the apprentice's violin to test it. If the apprentice has made a good violin, he is sent to another master. Otherwise, he is put back to work to gain more experience.

13 To get at the gist of the way style works, I've simplified the specific sociological claims. For more precise details, see, for example, W. Caudill and H. Weinstein, 'Maternal Care and Infant Behavior in Japan and America', *Readings in Child Behavior and Development*, in C. S. Lavatelli and F. Stendler (eds), New York, Harcourt Brace, 1972, pp. 78ff.

14 Deep Blue, the program that is currently world chess champion, is not an expert system operating with rules obtained from experts. Experts look at at most 200 possible moves, while Deep Blue uses brute force to look at a billion moves a second and

so can look at *all* moves seven moves into the future without needing to understand anything.

15 Yeats's last letter, in *The Letters of W. B. Yeats*, ed. Allen Wade, New York, Macmillan, 1955, p. 922, written just before his death, to Lady Elizabeth Pelham.

THREE DISEMBODIED TELEPRESENCE AND THE REMOTENESS OF THE REAL

1 E. M. Forster, 'The Machine Stops', *The New Collected Short Stories*, London, Sidgwick & Jackson, 1985. Written in 1909 partly as a rejoinder to H. G. Wells's glorification of science, 'The Machine Stops' is set in the far future, when mankind has come to depend on a worldwide machine for food and housing, communications and medical care. In return, humanity has abandoned the earth's surface for a life of isolation and immobility. Each person occupies a subterranean hexagonal cell where all bodily needs are met and where faith in the Machine is the chief spiritual prop. People rarely leave their rooms or meet face-to-face; instead they interact through a global web that is part of the Machine.

2 This sense of leaving behind one's body is also experienced when one does theoretical work. Descartes tells us that, in order to write his *Meditations*, he retired into a warm room where he would be free from passions and from having to act. Of course, one runs the risk that, from the detached, theoretical perspective, one may get a strange idea of what it is to be a human being, and, indeed, Descartes came to the conclusion that his body was not essential to him.

3 J. Mark 'Portrait of a Newer, Lonelier Crowd is Captured in an Internet Survey', *The New York Times*, February 16, 2000.

4 Ibid.

5 G. Johnson, *Wired Magazine*, January 2000.

6 Saint Augustine, *Confessions*, trans. R.S. Pine-Coffin, London, Penguin, 1961, p. 114.

7 I. Hacking, *Representing and Intervening*, Cambridge, Cambridge University Press, 1983, p. 194.

8 René Descartes, 'Dioptric', *Descartes: Philosophical Writings*, ed. and trans. Norman Kemp Smith, New York, Modern Library, 1958, p. 150.

9 Ibid., p. 235.

10 Ken Goldberg's famous piece of Web art, 'The Telegarden', is an example of such interaction at a distance. Visitors to this garden log in from terminals all over the world, directing a robot and camera to view, plant, and water seeds in a 6 ft. × 6 ft. patch of soil in a museum in Austria.

11 M. Merleau-Ponty, *Phenomenology of Perception*, trans. Colin Smith, London, Routledge & Kegan Paul, 1979, p. 302.

12 Ibid., p. 250.

13 This claim is argued for at length in Samuel Todes's *Body and World*, Cambridge, MA, MIT Press, 2001.

14 Merleau-Ponty, op. cit., p. 250.

15 R. M. Held and N. I. Durlach, 'Telepresence', *Presence*, vol. 1, pp. 109–11, as cited in Ken Goldberg (ed.), *The Robot in the Garden: Telerobotics and Telepistemology in the Age of the Internet*, Cambridge, MA, MIT Press, 2000.

16 J. Canny and E. Paulos, 'Tele-Embodiment and Shattered Presence: Reconstructing the Body for Online Interaction', in Goldberg (ed.), op. cit.

17 Personal communication.

18 M. Heidegger, *The Fundamental Concepts of Metaphysics*, trans. W. McNeil and N. Walker, Bloomington, IN, Indiana University Press, 1995, pp. 66–7.

19 W. H. Graves, '"Free Trade" in Higher Education: The Meta University', *Journal of Asynchronous Learning Networks*, vol. 1, Issue 1 – March 1997.

20 Merleau-Ponty, op. cit., p. 136.

21 That is, the player's gaze can't penetrate the distance to bring out more and more detail the way it does in real life. As Merleau-Ponty puts it: 'When, in a film, the camera is trained on an object and moves nearer to it to give a close-up view, we can *remember* that we are being shown the ashtray or an actor's hand, we do not actually identify it. This is because the screen has no horizons. In normal vision, on the other hand, I direct my gaze

upon a sector of the landscape, which comes to life and is disclosed, while the other objects recede into the periphery and become dormant, while, however, not ceasing to be there.' *Phenomenology of Perception*, 68.

22 Merleau-Ponty talks of the sense we have in the real world of there being no sharp boundary at the edge of our visual field, but, rather, of the world continuing behind our back. He further points out that if we felt the world behind us broke off suddenly, the scene in front of us would look different. 'The objects behind my back are . . . not represented to me by some operation of memory or judgment; they are present, they *count* for me. . . .' *Sense and Non-Sense*, 51.

23 Personal communication.

24 Forster, op. cit.

25 Experiments in computer-supported cooperation have shown that people are more inclined to defect in on-line communications than in face-to-face interactions, and that a preliminary direct acquaintance between people reduces this effect. So, computer technology can even weaken trust relationships already holding in human organizations and relations, and aggravate problems of deception and trust. See C. Castelfranchi and Y. H. Tan (eds), *Trust and Deception in Virtual Societies*, Dordrecht, Netherlands, Kluwer Academic Press, to appear 2001.

26 Yet people say MUD users fall in love in their chat rooms. I don't know what to make of that. Do they really trust each other, or, does such attraction perhaps show, as Shakespeare saw, that the erotic is more verbal than physical. (See, for example, Ulysses' description of the erotic attraction of Cressida, in *Troilus and Cressida*, IV, v, ll. 35–63.)

27 See, D. N. Stern, *The Interpersonal World of the Infant*, New York, Basic Books, 1985.

28 H. F. Harlow and R. R. Zimmerman, 'Affectional Responses in the Infant Monkey', *Science*, v, 130, 1959, pp. 421–32, H. F. Harlow and M. H. Harlow, 'Learning to Love', *American Scientist*, v, 54, 1966, pp. 244–72. In the experiment, an orphaned monkey was given two surrogate 'mothers' – a wire one and a terry-cloth one. To make the wire one more appealing, Harlow made the

feeding bottle part of the wire monkey. But in spite of this, whenever the small monkey was frightened, he would scurry to the terry-cloth monkey, not the wire one.

FOUR NIHILISM ON THE INFORMATION HIGHWAY

1 S. Kierkegaard, 'The Present Age', *A Literary Review*, trans. A. Hannay, London/New York, Penguin, forthcoming. Since Hanny's translation is not yet available, the page numbers in the text correspond to the Alexander Dru translation of *The Present Age*, Harper and Row, New York, 1962.
2 Ibid., p. 59.
3 S. Kierkegaard, *Journals and Papers*, ed. and trans. H. V. Hong and E. H. Hong, Bloomington, IN, Indiana University Press, vol. 2, no. 483.
4 Ibid., no. 2163.
5 Ibid.
6 J. Habermas, *The Structural Transformation of the Public Sphere*, Cambridge, MA, MIT Press, 1989.
7 Ibid., p. 94.
8 Ibid., p. 130.
9 Ibid., pp. 131, 133
10 Ibid., p. 138.
11 Ibid., p. 134.
12 Ibid., p. 137.
13 Kierkegaard, 'The Present Age', p. 62.
14 Ibid. pp. 62, 63 (My italics.)
15 Ibid., p. 77.
16 Ibid. p. 42.
17 Ibid., p. 68. (Kierkegaard's italics.)
18 Ibid., p. 77.
19 Kierkegaard, *Journals and Papers*, vol. 2, no. 480.
20 Ibid., no. 489. Kierkegaard would no doubt have been happy to transfer this motto to the Web, for just as no individual assumes responsibility for the consequences of the information in the press, no one assumes responsibility for even the accuracy of the information on the Web. Of course, no one really cares if it is reliable, since no one is going to act on it anyway. All

that matters is that everyone pass the word along by forwarding it to other users. The information has become so anonymous that no one knows or cares where it came from. Just to make sure no one can be held responsible, in the name of protecting privacy, ID codes are being developed that will assure that even the sender's address will remain secret. Kierkegaard could have been speaking of the Internet when he said of the Press: 'It is frightful that someone who is no one . . . can set any error into circulation with no thought of responsibility and with the aid of this dreadful disproportioned means of communication' (Kierkegaard, *Journals and Papers*, vol. 2, no. 481).

21 Kierkegaard, 'The Present Age', p. 64.

22 Although Kierkegaard does not mention it, what is striking about such interest groups is that no experience or skill is required to enter the conversation. Indeed, a serious danger of the public sphere, as illustrated on the Internet, is that it undermines expertise. As we saw in Chapter 2, acquiring a skill requires interpreting the situation as being of a sort that requires a certain action, taking that action, and learning from the results. As Kierkegaard understood, there is no way to gain practical wisdom other than by taking risky action and thereby experiencing both success and failure. Otherwise, the learner will be stuck at the level of competence and never achieve mastery. Thus the heroes of the public sphere who appear on serious radio and TV programmes, have a view on every issue, and can justify their view by appeal to abstract principles, but they do not have to act on the principles they defend and therefore lack the passionate perspective that alone can lead to egregious errors and surprising successes and so to the gradual acquisition of practical wisdom.

23 S. Kierkegaard, *Edifying Discourses*, ed. P. L. Holmer, New York, Harper Torchbooks, 1958, p. 256.

24 Ibid., p. 260.

25 Ibid., p. 262.

26 Kierkegaard, 'The Present Age', p. 103.

27 Ibid., p. 79.

28 Given Kierkegaard's use of the term 'sphere', then, precisely

because reflection is the opposite of taking any decisive action, and therefore the opposite of making anything absolute, what Habermas calls the public sphere is not a sphere at all.

A related non-sphere worth noting because it has become popular on the Net is Teilhard de Chardin's Noosphere, which has been embraced by the Extropians and others who hope that, thanks to the World Wide Web, our minds will one day leave behind our bodies. The Noosphere or mind sphere (in Ionian Greek 'noos' means 'mind') is supposed to be the convergence of all human beings in a single giant mental network that would surround the earth to control the planet's resources and shepherd a world of unified Love. According to Teilhard, this would be the Omega or End-Point of time.

From Kierkegaard's perspective, the Noosphere, where risky, embodied locality and individual commitment would have been replaced by safe and detached ubiquitous contemplation and love, would be a confused Christian version of the public sphere.

29 S. Turkle, *Life on the Screen: Identity in the Age of the Internet*, New York, Simon and Schuster, 1995, pp. 263–4.

30 Ibid., p. 180.

31 Ibid., p. 26.

32 A year after the publication of her book, Turkle seems to have had doubts about the value of such experiments. She notes that: 'Many of the people I have interviewed claim that virtual gender-swapping (pretending to be the opposite sex on the Internet) enables them to understand what it's like to be a person of the other gender, and I have no doubt that this is true, at least in part. But as I have listened to this boast, my mind has often travelled to my own experiences of living in a woman's body. These include worry about physical vulnerability, fears of unwanted pregnancy and infertility, fine-tuned decisions about how much make-up to wear to a job interview, and the difficulty of giving a professional seminar while doubled over with monthly cramps. Some knowledge is inherently experiential, dependent on physical sensations' (S. Turkle, 'Virtuality and its Discontents: Searching for

Community in Cyberspace', *The American Prospect*, no. 24, Winter 1996).

33 Kierkegaard, 'The Present Age', p. 68.

34 S. Kierkegaard, *Either/Or*, trans. D. F. Swenson and L. M. Swenson, Princeton, Princeton University Press, 1959, vol. II, pp. 16–17.

35 Ibid., vol. I, p. 46.

36 Ibid., vol. II, p. 197.

37 When I typed in Søren Kierkegaard, Google found 3,450 hits; Alta Vista found 7,452.

38 Kierkegaard, *Either/Or*, vol. II, p. 228.

39 J.-P. Sartre develops the idea of the absurdity of fully free choice in *Being and Nothingness*.

40 Sartre gives the example in *Being and Nothingness* of a gambler who, having freely decided in the evening that he will gamble no more, must, the next morning, freely decide whether to abide by his previous decision.

41 S. Kierkegaard, *The Sickness unto Death, A Christian Psychological Exposition for Edification and Awakening*, trans. A. Hannay, London/New York, Penguin, 1989, p. 100.

42 For Kierkegaard there are two forms of Christianity. One is Platonic and disembodied. It is expressed best in St Augustine. It amounts to giving up the hope of fulfilling one's desires in this life, and trusting in God to take care of one. Kierkegaard calls this Religiousness A, and says it is not the true meaning of Christianity. True Christianity, or Religiousness B, for Kierkegaard, is based on the Incarnation and consists in making an unconditional commitment to something finite, and having the faith-given courage to take the risks required by such a commitment. Such a committed life gives one a meaningful life in this world.

43 An attempt at inducing a sense of on-line risk was made in Ken Goldberg's telerobotic art project: *Legal Tender www.counterfeit.org*). Remote viewers were presented with a pair of purportedly authentic US $100 bills. After registering for a password sent to their e-mail address, participants were offered the opportunity to 'experiment' with the bills by burning or puncturing them at an on-line telerobotic laboratory. After choosing an experiment,

participants were reminded that it is a Federal crime to knowingly deface US currency, punishable by up to six months in prison. If, in spite of the threat of incarceration, participants click a button indicating that they 'accept responsibility', the remote experiment is performed and the results are shown. Finally, participants were asked if they believed the bills and the experiment were real. Almost all responded in the negative. So they either never believed the bills were real or else they were setting up an alibi if they were accused of defacing the bills. In either case, they hadn't experienced any risk and taken any responsibility after all.

44 As Turkle puts it: 'Instead of solving real problems – both personal and social – many of us appear to be choosing to invest ourselves in unreal places. Women and men tell me that the rooms and mazes on MUDs are safer than city streets, virtual sex is safer than sex anywhere, MUD friendships are more intense than real ones, and when things don't work out you can always leave' (S. Turkle, 'Virtuality and its Discontents: Searching for Community in Cyberspace', *The American Prospect*, no. 24, Winter 1996).

45 Kierkegaard, 'The Present Age', p. 80.

CONCLUSION

1 L. Guernsey, 'The Search Engine as Cyborg', *The New York Times*, June 29, 2000.

2 K. Chang, 'Science Times', *The New York Times*, September 12, 2000.

3 C. Thompson, 'Being There', *Fortune Magazine*, Special Issue on the Future of the Internet, 142: 8, October 2000, p. 236.

4 National Public Radio, *Talk of the Nation*, February 29, 2000.

5 H. Rheingold, *The Virtual Community: Homesteading on the Electronic Frontier*, rev. edn, Cambridge, MA, MIT Press, 2000.

6 Ibid., pp. 375, 376.

7 Ibid.

8 Ibid., p. 379.

9 Ibid., p. 382.

10 Ibid., p. 384.

11 Ibid.

12 F. Nietzsche, *Thus Spake Zarathustra*, trans. W. Kaufmann, New York, Viking Press, 1966, p. 34.

Index